Space Elevators are the Transportation Story of the 21st Century

Peter Swan, Ph.D.
Cathy Swan, Ph.D.
Michael Fitzgerald
Matthew Peet, Ph.D.
James Torla
Vern Hall

Prepared for the
International Space Elevator Consortium

July 2020

Space Elevators are the Transportation Story of the 21ˢᵗ Century

Copyright © 2020 by:

Peter Swan
International Space Elevator Consortium

Published by Lulu.com

info@isec.org

978-1-71674-663-5

Cover Illustrations*:*
Front – Amelia Stanton
Back – Peter Swan

Printed in the United States of America

Preface

A Network of Space Elevators enables humankind's movement off of Planet Earth

Elon Musk has stated he needs one million metric tonnes of supplies delivered to his colony on Mars.[1] In addition, the leadership of the Space Solar Power satellite constellation has stated they need five million metric tonnes delivered to Geosynchronous orbit.[2] Meanwhile, the European led Moon Village Association has stated they need "quite a lot of mass" delivered to the surface of the Moon to ensure a successful development for the gathering of people and missions. During this report, these three reference missions establish that the future needs of humanity demand more lift-off capability than is possible with rockets - alone. Therefore, operational collaboration between Galactic Harbours' Space Elevators and a variety of space launch competencies is essential. Together they will compose the movement to space during this century. That architecture enables movement of massive amounts of material for our journeys throughout the Solar System.

This monumental and critical story is still being written. The Apex Anchor

is where a Galactic Harbour meets the Shoreline of Outer Space -

where the "Transportation Story of the 21st Century" meets the "Final Frontier."

---

[1] Elon Musk, 21 July 2019 CBS Sunday Morning Interview and McFall-Johnsen, M. Dave Mosher, "Starship rockets every day and creating a lot of jobs on the red planet," Business Insider, Jan 17, 2020.

[2] Mankins, John, personal conversation with P. Swan, at IAC, Washington.D.C. Oct 2019.

Acknowledgements

The authors thank the active members of the International Space Elevator Consortium, and other Space Elevator enthusiasts, who have been dreaming "big" for years. The remarkably creative analyses accomplished over the last ten years have allowed the Space Elevator body of knowledge to increase exponentially. This latest contribution will expand the vision and explain why Space Elevators are essential to movement off-planet as a permanent transportation infrastructure. Each of us knows that when Space Elevators are operational, movement off-planet proceeds robustly.

Thanks are also due to those who have contributed during the extensive review - especially Dr. Cathy Swan who read every word - several times - as an editor translating engineering into English. In addition, thanks to co-authors Vern Hall and Michael Fitzgerald for their significant insight driving forward the concept of a permanent transportation infrastructure.

And of course, thanks must go to the innovative Arizona State University research team. The two leads were Matthew Peet, Ph.D., Associate Professor, and James Torla, student leader conducting critical simulations with remarkable results. Their team of students included Mark Lyons, Runa Nakamura, Renzo Curay De La Ros, Shawn Michael Bauer, Nathan Renard, Nicholas Iannacone, Tyler Mebane, Avi Brahmbhatt, Samuel Bolar, Jose Valenzuela, Jonathan Johnson, and Ryley Miller. This orbit-breaking research at ASU has lead to new concepts and strengths of space elevators inside the interplanetary mission arena.

And of course, thanks and recognition must go to the "Peer Review Team" that helped the final presentation of the whole document: John Knapman, Dennis Wright, and Paul Phister.

We believe we will have a huge impact on the future and our ability to move humanity forward. This Massive Transformative Moonshot is rewarding to work on and is being accomplished because we believe in our mission.

Well done Space Elevator team!

Peter A. Swan, Ph.D.
President, ISEC
Study Team Lead

Executive Summary

The International Space Elevator Consortium's (ISEC) research of customer needs, and Space Elevators' capabilities to satisfy them, has expanded greatly - to include:

There needs to be a <u>Change of Vision for Interplanetary Movement</u> when delivery of mass is inexpensive, timely, routine, environmentally friendly, daily, and supportive. It turns out that projections of transportation capabilities of Space Elevators provide immense opportunities and ensure that humanity can "bring with them" the essential elements for survival and aggressive growth. A new vision of Galactic Harbour architectures (growth from single Space Elevator to a pair working together) will change the thinking for off-planet migration - We <u>can</u> bring it with us!

Movement off-planet will <u>require complementary capabilities</u> such as rocket portals and Galactic Harbour infrastructures, each with their own strengths. Rockets have the capability to move people through radiation belts rapidly. This is compatible with Space Elevators' capability of massive movement in a timely, routine, inexpensive and Earth friendly manner.

<u>Expectations by the user for massive cargo throughput:</u> Comparing estimated customer needs ("Demand Pull") for three reference missions (Space Solar Power, Colonies on Mars, and Lunar Villages) leads to the conclusion that Space Elevator Infrastructures are required to fulfill future missions.   These permanent transportation infrastructures enable movement off-planet:
- Fast Transit (as low as 61 days to Mars)
- Daily departures to Mars (no 26 month launch window queue)
- Massive lift-off capacity, at full operational maturity, of 170,000 metric tonnes per year towards GEO, the Moon, Mars, and beyond.

After the study team looked at these phenomenal results, and gained new insights into travel to other planets (and the Moon), the team recognized that they have revolutionized the approach for movement off-planet. The change in mindset is basic. Space Elevators CAN ship essential cargo from Earth-based suppliers. The travelers no longer have to establish supply sources beyond GEO (such as immediately building infrastructures on the Moon - all from regolith). Mass is Good (an anathema for rocket supply concepts)! Two statements, supported in this study report, must be shared across space faring nations:

The Space Elevator is closer than you think!   &
Space Elevators enable off-planet movement!

Table of Contents

The Awakening

A NEW Space Transportation Paradigm has emerged. Ideas brought forward in this report are presented in clear and understandable ways, although a revolutionary concept is becoming real. The envisioned cooperative and collaborative operations between rockets and space elevators will benefit mankind into the next century. This Space Elevator architecture, as a remarkable Earth to Space Transportation infrastructure, provides logistics support to future missions throughout our solar system. This is a path to fully support interplanetary travel occurring by the second half of this century. This report will highlight the following realizations:

- Space Elevators can be accomplished because there is a material for the tether.
- Space Elevators ENABLE Interplanetary Missions
    - Fast Transit to Mars (as short as 61 days, with variations out to 400+)
    - Can release towards Mars EVERY day (no 26 month wait)
    - Can move massive amounts of cargo (170,000 Metric tonnes/year to GEO and beyond)
- Space Elevators are Earth Friendly
    - Space Solar Power replaces 100s of coal power plants
    - No rocket exhaust to contribute to global warming
    - No additional space debris
    - Opens up remarkable commercial enterprises at Earth Ports, GEO Regions and beyond
- Offer to all future scientists
    - Any size science experiment
    - Any solar system destination
    - Releases every day towards multiple research destinations

How can this be possible? Simple - a working Space Elevator defeats gravity and the traditional rocket equation. Massive payloads to the Apex Anchor[3] - raised by electricity - to be released at 7.76 km/sec towards destinations; daily, routinely, safely, and robustly all while being environmentally friendly.

---

[3] Apex Anchor is term for smart counterweight at approximately 100,00 km altitude.

Chapter 1 - Introduction - Enabling Interplanetary Missions

Galactic Harbours will be at the center of the transportation story of the 21st century.

1.0 The Galactic Harbour:

The Galactic Harbour is the unification of Transportation and Enterprise. From an engineering aspect, the Galactic Harbour is the combination of the Space Elevator Transportation System and the Space Elevator Enterprise System. The Galactic Harbour will be the volume encompassing the Earth Port while stretching up in a cylindrical shape to include two Space Elevator tethers outwards beyond the Apex Anchor.

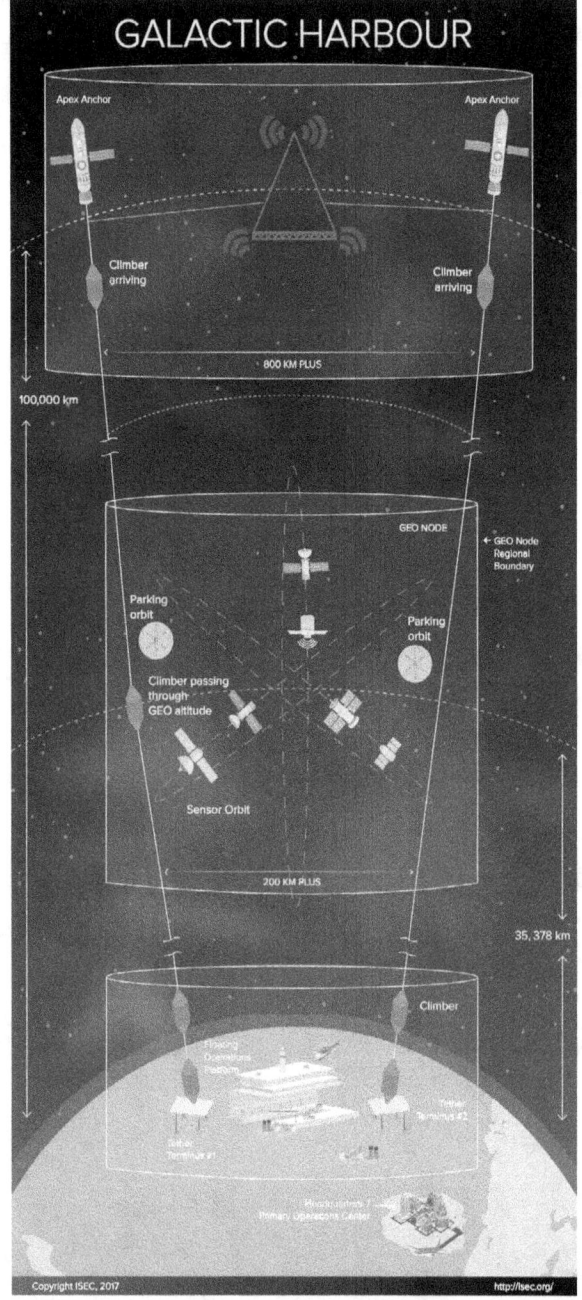

Figure 1.1, Galactic Harbour

Customer product/payloads will enter the Galactic Harbour at the Earth Port and exit someplace up the Tether. There will be tremendous enterprise development in the GEO Region; such as spacecraft assembly, refueling operational satellites, solar power collection and, of course, businesses will emerge supporting flight operations from the Apex Anchor; trips to interplanetary destinations. From an operational aspect, The Transportation System is the "main channel" in the Galactic Harbour, moving cargo from the Earth Port to the transportation locations within the Harbour; i.e. the GEO Region and the Apex Anchor Region. The new businesses at GEO, at the Apex, and the businesses at the Earth Port are the Space Elevator Enterprise system.

When the Elevator becomes operational, it will service all of these enterprises. With this concept of Galactic Harbours comes the recognition that movement off-planet will require complementary capabilities, such as rockets portals and Galactic Harbour infrastructures, each with their own strengths and short-falls.

1.1 Introduction to Movement off-Planet:

As humanity expands off Earth, the need for support increases exponentially. The mass per day required to be delivered to the Moon, Mars and other destinations will out stretch the current rocket based approach -- something revolutionary is required. The authors project into the future and see that Space Elevators allow the growth of humans off-planet to accelerate with three major improvements: - massive movement of mission support equipment - a tremendous opening up of launch windows - shorter travel times. This view of the future is encouraged by national efforts around the world, to include aggressive approach towards the Moon by the United States. Not only has the government committed to going back to the Moon, but stated that they will have female boots on the Moon to match the historic ones that led the way. Space Policy Directive-1 (Dec 2017) stated, "Lead an innovative and sustainable program of exploration with commercial and international partners to enable human expansion across the solar system." (Appendix H expands on this)

With Mars as the destination of a recent Arizona State University (ASU) research study[4], the results showed the delivery time for supplies from the Earth can become very short using Space Elevators. In addition, the concept of one launch window every two years towards Mars is collapsed to multiple launches each week. The essence of this change is the daily release with a tripling of kinetic energy at the Earth's sphere of influence (SOI) [compared to rockets entering the Hohmann transfer ellipse]. The resultant increase in energy is enabled by the release from a 100,000 km altitude Apex Anchor rotating with the Earth. These potential and kinetic energies result in a hyperbolic orbit departing the Earth. This ASU study was aimed at determining "time of flight" from Earth to Mars when departure angle and energy at the edge of the SOI reflected the strengths of having Space Elevators at the Earth's equator. This study report expands upon the recognition of daily departures to Mars with tremendous energy and addresses the opportunities available for massive logistics support. This report will focus upon three Apex Anchor characteristic strengths not exploited before. They are summarized as:

- Fast Transit to destination (Mars as short as 61 days). Arizona State University (ASU) research into release from an Apex Anchor with the concept of a Lambert

---

[4] Torla, James, Matthew Peet, Optimization of Low Fuel and Time-Critical Interplanetary Transfers using Space Elevator Apex Anchor Release: Mars, Jupiter and Saturn. IAC-19, paper and presentation, Washington D.C., Oct 2019

Problem solution shows remarkable transit times periodically during the 26 month repeating orbital relationship between Earth and Mars.

- Massive liftoff capability (14 metric tonnes payload per day, initial capability). Space Elevators start out with huge throughput capacity with daily liftoffs (5,110 metric tonnes / year). In addition, there will be remarkable growth as the tether material and infrastructures mature. The Initial Operational Capability starts at 14 metric tonnes of payload per day with the Full Operational Capability reaching 79 metric tonnes.
- Daily departures available (no waiting for 26 month Mars Launch Windows). The ability to launch each day towards Mars is a revolutionary concept vs. the traditional wait period of 26 months (the dreaded 26 month launch window restrictions currently in place for rockets). Transit times for cargo can vary over the repeating planetary dance; but, they can be started towards Mars each day simplifying the mission support concept.

To accomplish these three topics within the study, the team assessed many items. The study report discusses significant findings in each of the following study topics:

- Change of Vision for Interplanetary Movement when delivery of mass is inexpensive, timely and supportive. It turns out that the new revelations in transportation capabilities of Galactic Harbours open up immense possibilities and ensure that humanity can "bring with them" the essential elements for survival and aggressive growth. This new vision of Galactic Harbour architectures will change the "thinking" for off-planet migration - We can bring it with us!
- Movement off-planet will require complementary capabilities, such as rockets portals and Galactic Harbour infrastructures, each with their own strengths and short-falls. Moving people through the radiation belts rapidly is a strength of rockets while massive movement in a timely, routine, inexpensive and Earth friendly manner are the strengths of Space Elevators.
- Discussion of mission needs when analyzing mass to location. This discussion will start the analysis of "how much carrying capability" is required by each supportive infrastructure; when, to where, and their priority. In the past, the rocket approach valued light-weight compact designs of support equipment while the space elevator permanent infrastructure will enable mass to be moved to the desired destinations easily. In addition, volume restrictions are greatly reduced. The driving function becomes needs of the customer, not light-weight designs.

After the study team looked at these phenomenal results, and gained new insights into travel to other planets (and the Moon), the team recognized that they have changed the approach for movement off-planet. The change in mindset is basic - We CAN ship essential cargo from Earth-based suppliers. We no longer have to establish supply sources beyond GEO (such as immediately building infrastructure on the Moon, all from regolith). Mass is Good! (an anathema for rocket supply concepts)

Figure 1.2, Newly Recognized Strengths

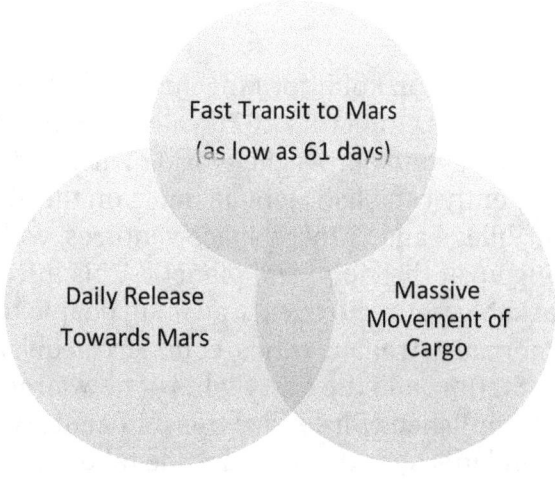

Meanwhile, to place the Galactic Harbour within the future movement off-planet, the concept of a Mosaic of Space should be developed. When one looks at missions for Space Elevators within the Galactic Harbours of the future, one sees a tremendous diversity of roles. This Mosaic of Space concept has been portrayed as a combination of trade, exploration, enterprise, research, and government. As such, multiple Galactic Harbours around the world will enable all sorts of activities and become catalysts for change. The essential value of low cost access to space – for all humankind – is brought forward by the Space Elevator Transportation System. This is accomplished through delivery of payloads beyond Earth's gravitational influence - routinely and daily. This permanent transportation infrastructure will complement rockets of the future in missions as essential partners. Two Space Elevator regions are evident and are discussed extensively in this report. The GEO Region is the destination for an extensive variety of entrepreneurial thrusts. One thrust will be the classic view: refuel, repair, upgrade, or replace the fleet of space flight systems milling around GEO. Another thrust will be the factory operations in this new industrial GEO region and all that goes with it. This second thrust will eventually include research and testing of new yet unimagined industries that operate in low gravity. One principal mission of tremendous significance to the Earth's population is the development and operation of space solar power - essentially replacing coal power plants with electrical power transmission to the Earth's surface from GEO. The second destination, Apex Anchor Region, is the location for interplanetary efforts - in fact, all interplanetary efforts. Departures and arrivals from Earth will be from Apex Anchors that serve as a launching pad already moving at over seven kilometers per second toward all interplanetary destinations. A Mosaic of Space concept recognizes remarkable diversity of missions to be accomplished from Space Elevator transportation infrastructures in conjunction with remarkable business enterprises across the various regions of Galactic Harbours. The study results from Arizona Statue University have changed the approach for interplanetary missions. Support to those missions can be timely and enable massive cargo demands on a much smaller budget and with much greater frequency. Indeed, the Galactic Harbour will be instrumental in opening up space to all mankind.

1.2 Background:

1.2.1   Inflection Point for Movement Off-Planet: This exciting future is based upon the realization that 2020 will be an inflection point in growth towards space activities beyond our fragile world. The remarkable moves by the United States to permanently establish "female boots on the Moon," tied together with the European Lunar Village and Chinese space ventures, will set the stage for hope of a future looking up at the Moon and planets. This third decade of this millennium will have remarkably powerful events pushing people towards challenging goals such as commercial astronaut wings, commercial delivery of astronauts to the International Space Station, and the first flights of new large reusable rockets. What is expected to be accomplished will service this mid-century expansion off-planet. Hang on to your hats (and ideas) as this new decade takes off.

1.2.2   Enhanced Needs at Inflection Point: The Galactic Harbour architecture is in the center of the transportation story of the 21st century. Reliable, routine, safe, environmentally friendly, and efficient access to space is close at hand. Space Elevators are the key elements of Galactic Harbours, and an essential part of the global and interplanetary transportation infrastructure. In the community of off-planet movement with NASA's newest move to put boots on the Moon by 2024, Space Elevators must be included in the discussions. The key to any support infrastructure is daily, routine, inexpensive, environmentally friendly, and massive movement of payloads towards the Moon and Mars. These are strengths of Apex Anchors - especially with high velocity and daily launch windows. This study provides impetus for discussions about the three major attributes when departing from the Apex Anchor and the other strengths of Space Elevators with respect to future missions to GEO, Moon, Mars, asteroids and beyond. To a large extent, in the orbital characteristics area, this study will leverage research accomplished at Arizona State University on the Lambert Orbital Solution and fast transits on those direct elliptical orbits to Mars and other celestial destinations. (see Chapter 4 on ASU Research results)

The reality is that when humanity decides to go to the Moon and Mars, there will be a tremendous need for logistics support [mission support equipment] as well as transporting people [especially at low cost, routinely and daily]. Mission logistics support describes all the extra mass that needs to be delivered to their destinations. For this study, rockets will move people well past the Space Elevator Initial Operational Capability (IOC). Movement of people on Space Elevators will arrive in time, but well after the IOC architecture (roughly 2036). The logistics support needed by humans beyond Low Earth Orbit starts with everything required to stay alive and enable mission success. Much will accompany flights of humans to the Moon and Mars. However, mission support equipment will be required to enable humans to prosper with pre-positioning of supplies and equipment; equipment that accompanies the human missions; and, post arrival mission survival and enhancement equipment and supplies. Indeed, the travelers will learn to live off the land (in-situ resources). However, humans will always need enormous quantities of support infrastructure delivered to wherever they go. The questions to ask are: How do the strengths of Galactic Harbours enable missions to Moon and Mars? Can a release from the Apex Anchor of 14 metric tonnes daily, routinely and safely towards Mars (and/or the Moon) be supported as an amazing and required capability? Missions to Mars need to be analyzed to show non-traditional (non-Hohmann Transfer) orbital options with high energy, fast transit, characteristics. The puzzle is how to best support Mars missions with Galactic Harbour's tremendous strengths of daily release and massive movement. One interesting twist within this analysis is that Elon Musk recently established a baseline of needs on Mars by mid-century.

One Million Tonnes to Mars to Support my Colony![5]

Is it a coincidence that Mr. Musk asked for delivery of a million tonnes of support infrastructure for his colony on Mars? I think not! It is obvious to anyone studying recent commercial off-planet movement that there is a crying need for routine, daily, inexpensive, environmentally friendly, safe, and massive payload capability access to space. It just so happens that Space Elevators, within Galactic Harbours, have those strengths. The International Space Elevator Consortium (ISEC) believes in these missions of mid-21st Century:

Galactic Harbours Enable Interplanetary Mission Support.

This study also projects mission needs to support both a Lunar Village and Space Solar Power satellite infrastructures at GEO (supplying inexpensive electrical power to the surface of the Earth). This year's study activity will place the interplanetary mission needs in the perspective of other Space Elevator missions.

1.2.3 The time is NOW for Galactic Harbours: There has been great progress in the last ten years - we are definitely 20 years ahead of the original "doable" Space Elevator of Dr. Brad Edwards (2002). ISEC has taken the concept way beyond what he initiated. ISEC believes we have Space Elevators to the point where "They should be included at the Big Table during discussions of future needs in space (especially beyond LEO)." We believe we are ready to initiate a Moonshot - build a transportation infrastructure that will benefit humanity with revolutionary low-cost access, routine daily liftoffs, safe rides with remarkable support for interplanetary missions (daily launch to Mars - without 26 month launch window restrictions with massive payloads on a fast track delivery in as little as 61 days). ISEC is ready to proceed by bringing in engineering firms to develop programs and begin testing of major segments of Space Elevators, within Galactic Harbours.

---

[5] Elon Musk, 21 July 2019 CBS Sunday Morning Interview and McFall-Johnsen, M. Dave Mosher, "Starship rockets every day and creating a lot of jobs on the red planet," Business Insider, Jan 17, 2020.

In addition, if one were to look at the global approach, the Japanese and the Chinese are pursuing Space Elevators. The Japanese are very open about what they are doing with their government funding while the information on Chinese plans are less overt. Space Elevators, in the Japanese culture, have an established history and could be an excellent strategic asset for their country. The transportation aspect of this permanent infrastructure for access to space is core to understanding their overall needs and benefits. The key to this whole discussion is that (1) the Japanese popular culture understands that Space Elevators will be there in the future for liftoffs from Earth, (2) their research organizations have delved into the details of Space Elevators and understand them, and (3) the Obayashi Corporation plans on building Space Elevators. The Japanese need power for their island nation and expect it to come from space solar power, by 2050. This combination of these realities could lead to massive Space Elevators for low cost, permanent infrastructures and environmentally friendly access to space. China has recently envisioned itself as a global economic leader. It certainly is in the Western Pacific region. As a "spacefaring" nation, it had a late start; but recently, it has taken aggressive steps. It has a rover on the Moon. It has (or soon will have) its own satellite navigation system. It has been very aggressive "interacting" with other nations along the GEO belt. In the last ten years, a new Chinese space vector has emerged fueled by academic and scientific thrusts and has made strong efforts to cultivate commercial space entrepreneurial positions. These common elements have melded with commercial space efforts around the world. China's macro trade strategy (called the Belt and Road Initiative) includes their commercial space initiatives, their announced Chinese Space Elevator, and Chinese space control activities.

The time is now for the development of commercial efforts to build the first Space Elevators. The need is now visible to support movement off-planet and significant missions to GEO, such as space solar power. Understanding of engineering development is clear and now has a recognized material available for the tether. Development of Galactic Harbours should be initiated now!

1.2.4 The 7th Architecture - Standing on the shoulders of others: If you have been around awhile, you would have seen that much has changed across the topic of access to space during this new century. So, where did the enthusiasm for Space Elevators really start? In this case, much of it was started by others; and grown incrementally. The current participants stand with this legacy. The 7th Space Elevator Architecture, as described in this study report (explained in an appendix) was first presented to a small but enthusiastic group of attendees at the International Space Development Conference (ISDC) in June 2019.

This "passing of the baton" occurred during the ISDC Space Elevator Track while Jerome Pearson (Space Elevator "co-inventor"), Michael Fitzgerald (ISEC "Chief Architect") and a generation of "excited students" participated. This was more than an assembly of young and old. It was also a portrait of the stewards of the Space Elevator revolution -- from Inventor to Developer to Innovators.

Figure 1.3, Passing Baton (ltr, Mukherjee, Pearson, Fitzgerald, Dressler)

Figure 1.3 shows Souvik Mukherjee (16 year old) with panel members. Figure 1.4 shows James Torla (ASU student researcher) presenting his research results during the afternoon. The 7th architecture lives on the shoulders of previous inventors and engineers while it looks into the future along with the tremendous efforts to move off-planet.

What entails a Space Elevator Architecture? Key features leading to the list below include comparison with significant published descriptions of Space Elevators separating out each based upon a few important criteria: (1) publishing, and distribution of concepts, (2) the engineering level of detail, and, (3) presentations showing as much of the available engineering as possible at that time. Each concept developed credibility from its own insights while ensuring movement forward for the

development of Space Elevators. This led to a series of seven architectures over the last **125 years.** The first two were significant leaps in understanding while the next four have enabled the current status of readiness for development. The recent growth to a dual Space Elevator Galactic Harbour Architecture leads to the conclusions from this report that they ENABLE interplanetary mission support, leading to robust movement off-planet.

Figure 1.4, James Torla at ISDC 2019

19th Century, The first Architecture
- In 1895, Tsiolkovsky looked at the Eiffel Tower and envisioned something built up towards Geosynchronous Orbit. As the innovative thinker across the turn of the century, he saw many future activities in space with unique characteristics - one of which was to "climb to space."

20th Century Architectures
- In 1960, Yuri Arsutanov presented a realistic approach visualizing how to achieve Mr. Tsiolkovsky's vision - base a string centered at geosynchronous altitude reaching from the ocean to deep in space.
- In 1974, Jerome Pearson resolved many issues with engineering calculations of tether strengths needed and approaches for deployment. This was once again a leap beyond Mr. Arsutanov's work and set the stage for the "modern design" for Space Elevators.

21st Century Architectures - Modern Concepts
- Dr. Brad Edwards established the initial baseline for Space Elevator infrastructures at the turn of the century with his book: "Space Elevators" [2002]. He established that the engineering could be accomplished in a reasonable time with reasonable resources
- In 2014, The International Academy of Astronautics (IAA) leveraged Dr. Edwards' design supported by intervening ten years of excellent development work from around the globe. The 41 space experts combined to improve the concept and establish new approaches.
- In parallel during 2014, a new version of Space Elevator architectures was released by the Obayashi Corporation. Their set of assumptions for their study established stricter requirements and resulted in longer developments with increased payload capacity. Their focus was movement of humans and massive loads to GEO and beyond.
- **A Seventh Architecture:** In 2019, the Galactic Harbour Architecture was shown in the IAA latest study report.[6] The concept was later developed and matured within the ISEC study reports showing progress.[7] The current version of the individual Space Elevator incorporates the ideas expressed within this IAA study, "Road to the Space Elevator Era." It was completed with inputs from over 47 space experts and leaped ahead of the previous architectures to include Galactic Harbours, solar energy to drive the tether climbers, and other important additions to the concept. It also recognized that the tether material has moved on from carbon nanotubes towards new revolutionary 2D materials such as single crystal graphene or single crystal hexagonal boron nitride (hBN). Galactic Harbours show major strengths through the combination of the necessary transportation infrastructure with commercial enterprises which will develop naturally within. The resulting

[6] Swan, P., David Raitt, John Knapman, Akira Tsuchida, Michael Fitzgerald, Yoji Ishikawa, Road to the Space Elevator Era, **Virginia Edition Publishing Company**, Science Deck (2019) ISBN-19: 978-0-9913370-3-3

[7] Swan, Peter, Michael Fitzgerald, "Today's Space Elevator," ISEC Study Report, lulu.com, 2019.

vision of Galactic Harbours (see chapter two in this report) show multiple locations around the equator leading to six or more Space Elevators inside three Galactic Harbours supporting, as a principal mission, interplanetary logistical support.

"What is past is prologue" certainly applies to what happened here. However, the authors of this work see it more as their stewardship built on discreet steps; an improving and enlarging legacy. Standing on their shoulders we unabashedly proclaim: "We have moved from Vision to Necessity."[8]

1.2.5 Lexicon of Terms: In the lexicon for Space Elevators (see Appendix G), the following are in play for this study report, consistent with global understanding of the terms:

- Space Elevator (SE) - single tether of 100,000 km length
- Initial Operational Capability (IOC) - estimated capacity of 14 Metric tonnes of cargo each day.
- Full Operational capability (FOC) - estimated capacity of 79 Metric tonnes of cargo per day for a mature Space Elevator with human passengers as well as cargo.
- Galactic Harbour (GH) - Transportation Infrastructure with robust enterprises along the 100,000 km dual Space Elevators.
- At the Horizon - three IOC Galactic Harbours
- Beyond the Horizon - three FOC Galactic Harbours growing to many around the equator.

1.3 Space Elevator Developmental Status:

A "Sea State Change" has occurred within developmental progress of the Space Elevator concept. Significant activities have occurred around the world, as reflected in the IAA 2019 study report "A Road to the Space Elevator Era."[9] These engineering activities have led to a successful Preliminary Technological Assessment. This surfaced after multiple organizations accomplished major tasks:
- ISEC produced ten year-long studies with resulting reports (Appendix D).
- The International Academy of Astronautics produced two study reports supporting the concept (Appendix E).
- The Obayashi Corporation conducted an independent study that focused upon humans on the Space Elevator and massive movement of space based solar power satellites to GEO.[10]

---

[8] Fitzgerald, M., words from an ISEC memo 2020.

[9] Swan, P., David Raitt, John Knapman, Akira Tsuchida, Michael Fitzgerald, Yoji Ishikawa, Road to the Space Elevator Era, **Virginia Edition Publishing Company**, Science Deck (2019) ISBN-19: 978-0-9913370-3-3

[10] Ishikawa, Yoji, The Space Elevator Construction Concept, Obayashi Corporation, 2013, IAC-13-D4.3.6.

- Internal ISEC assessments were provided by a series of ISEC Chief Architect's Notes. (see www.isec.org).
- The agendas of major international space agencies are aligning to target human presence and/or settlements on the Moon and Mars, thereby establishing demand pull.

Because of these achievements, there are many questions that have been answered over the last ten years with respect to the readiness of Space Elevators. The position of ISEC is that the:

> *Space Elevator development has gone beyond a preliminary technology readiness assessment and is ready to enter initial engineering validation testing -- leading to the establishment of needed capabilities.*[11]

With this statement, there are many questions that are still of interest. As such, in Appendix C, the authors have answered the questions shown next. As this is a detour off the main theme of the study, the multiple page answers are shown in the Appendix.   These questions have been answered around the community with significant contributions from ISEC and the International Academy of Astronautics.

- What is the current schedule as seen by IAA and ISEC?
- Where is the development of the Space Elevator inside Engineering Stages?
- What is the material for the tether and where is it in its development?
- How will the Multi-Stage Space Elevator concept enable early start?
- Do we have a "gold standard" for simulations?
- Assured Space Debris Survivability?
- What are the near term activities that must be initiated?

1.4 Unique Strengths for Interplanetary Flights:

During last year's research at ASU, some major findings were developed that will enable robust interplanetary missions (expanded in Chapter 4, ASU Research). Galactic Harbours will be spread out along the equator during the late 2030's with an estimate of three being operational by 2040. Each Galactic Harbour will have a pair of Space Elevators. This study report looks at three unique characteristics that are inherent to a permanent infrastructure.

---

[11] Swan, Peter, Michael Fitzgerald, "Today's Space Elevator," ISEC Study Report, lulu.com, 2019.

1.4.1 Major Finding #1 - Fast Transits to Mars Available: With the daily release of payloads towards Mars (and other interplanetary destinations), the Apex Anchor imparts tremendous velocity with very little drag from Earth's gravity. As a result of studies conducted by Arizona State University (led by James Torla and Associate Professor Matthew Peet), a periodic fast transit to Mars lowers this time to 61 days. This occurs frequently with respect to the planet's orbital positions; but, it is still 61 days to Mars a few times every 26 months. This research and analysis is expanded and discussed in Chapter 4. Figure 1.5, Fast Transit - 76 Days, shows the non-Hohmann Transfer approach from Earth (smaller circle) to Mars (larger circle) in an optimum orbit along an ellipse of immense energy. The ellipse can "cut the corner" because it has high velocity.

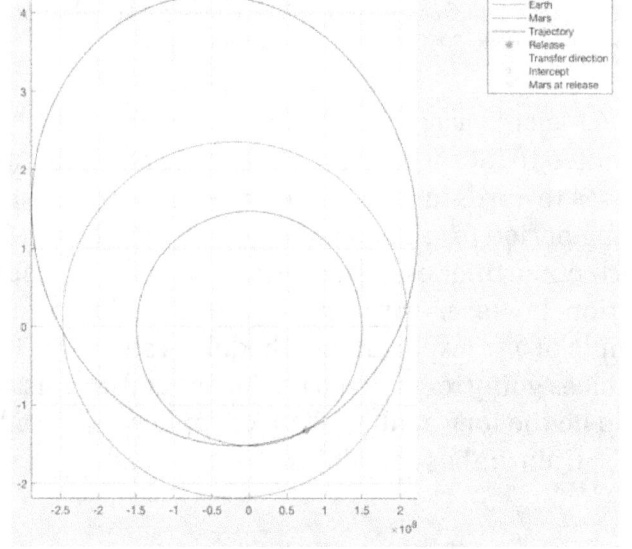

Figure 1.5, Non-Hohmann Transfer Fast Transit, Earth to Mars[12]

1.4.2   Major Finding #2 - Routine Massive Lifts: Initially, each Space Elevator Tether Climber will carry 14 metric tonnes of payload to GEO and beyond with departures scheduled every day. (or 5110 Metric tonnes per year). Once the three Galactic Harbours are established, with two Space Elevators each, the yearly throughput would be 30,660 Metric tonnes. Each tether capacity would probably double every few years. Before long, the carrying capability of Galactic Harbours would be roughly 100,000 Metric tonnes per year. A current question being asked is how many rocket launches are required to support movement off planet? Compare that question with the next table that shows an estimate of lift capability over a few years (see chapter 5 for explanation of Space Elevator Throughput). These numbers revolutionize the concept of interplanetary mission support. When you need a piece of equipment or a habitat, it can be delivered from a routine, daily, safe and timely transportation infrastructure from the Earth to its destination.

---

[12] Torla, James and Matthew Peet, Optimization of Low Fuel and Time-Critical Interplanetary Transfers using Space Elevator Apex Anchor Release: Mars, Jupiter and Saturn, IAC-19, paper and presentation, Washington D.C., Oct 2019.

Table 1.1: Number of Rocket Launches vs. Galactic Harbour Lift-Offs

Type	Lift Average (Metric tonnes)	Number per year	Total Mass to Interplanetary (Metric Tonnes) per year
Individual Heavy Launch Vehicles	10 per launch	87 (average last 5 years)	870 tonnes if all went to this single mission
Individual Heavy Launch Vehicles - SpaceX StarShip (estimate 2025)	100 per launch	100 times per year	10,000 tonnes
Galactic Harbour Transportation Infrastructure at Initial Operational Capability (estimate 2040)	6 tethers x 14 tonnes = 84 tonnes per day	every day towards Mars and Moon	84 x 365 = 30,660 tonnes
Galactic Harbour Transportation Infrastructure at Full Operational Capability (estimate 2050)	6 tethers x 79 tonnes = 474 tonnes per day	every day towards Mars and Moon	474 x 365 = 173,010 tonnes

1.4.3 Major Finding #3 - Routine Daily Lifts: As Space Elevators are scheduled to lift cargo daily, the release from the Apex Anchor is parallel. Daily releases towards interplanetary missions will be standard. This means there is no waiting period for a launch window towards Mars (currently it is 26 months between launch opportunities). The variation in flight paths (orbits) and time will range in duration; however, cargo can now be scheduled for launch every day of the year. An example of a transportation schedule is shown in Table 1.2. This table is an example of choices going to Mars during July of 2035. (It is an exemplar, as showing every day would be too long a table.) Think supplies delivered by infrastructures similar to FedEx, train, or bus schedules.

Table 1.2, Bus Schedule to Mars from Apex Anchor

1.5 Basic Strengths of Space Elevators and Galactic Harbours:

The baseline Initial Operations Capability (IOC) Space Elevators will not be carrying people; so, their missions are strictly logistics support to GEO and beyond - including interplanetary. If you think in today's world for a parallel transportation infrastructure example, such as planes for people and rail or sea for logistics, you can begin to understand the strengths of Galactic Harbours for support to interplanetary

This is the transportation story of the 21st century. Reliable, safe, and efficient access to space is close at hand. The Space Elevator is the Galactic Harbour, and an essential part of the global and interplanetary transportation infrastructure.

Bus Schedule for Interplantary Transportation
when departing from Galactic Harbour Apex Anchor

Bus Schedule, from Apex Anchor 2035					
Date	Departure	Destination	Flight Time	Arrival	Comments
7/1/2035	Indian #1	Mars	87 days	9/26/2035	
7/1/2035	Pacific #1	Mars	86 days	9/25/2035	
7/1/2035	Pacific #2	Mars	84 days	9/22/2035	Fast

Bus Schedule, from Apex Anchor 2035					
Date	Departure	Destination	Flight Time	Arrival	Comments
7/8/2035	Indian #1	Mars	81 days	4/14/2035	
7/8/2035	Indian #2	Mars	81 days	4/14/2035	
7/8/2035	Indian #1	Mars	80 days	4/13/2035	Fast

Bus Schedule, from Apex Anchor 2035					
Date	Departure	Destination	Flight Time	Arrival	Comments
7/15/2035	Indian #1	Mars	79 days	10/2/2035	
7/15/2035	Indian #1	Mars	79 days	10/2/2035	
7/15/2035	Indian #2	Mars	79 days	10/1/2035	
7/15/2035	Indian #2	Mars	79 days	10/1/2035	
7/15/2035	Pacific #1	Mars	78 days	9/30/2035	Fast
7/15/2035	Atlantic #1	Mars	190 days	1/21/2036	
7/15/2035	Atlantic #1	Mars	182 days	1/13/2036	
7/15/2035	Atlantic #2	Mars	173 days	1/4/2036	
7/15/2035	Atlantic #2	Mars	164 days	12/25/2035	
7/15/2035	Atlantic #1	Mars	154 days	12/15/2035	

Bus Schedule, from Apex Anchor 2035					
Date	Departure	Destination	Flight Time	Arrival	Comments
7/22/2035	Pacific #2	Mars	77 days	10/7/2035	Fastest
7/22/2035	Pacific #2	Mars	77 days	10/7/2035	Fastest
7/22/2035	Pacific #1	Mars	223 days	3/1/2036	

Bus Schedule, from Apex Anchor 2035 to Moon					
Date	Departure	Destination	Flight Time	Arrival	Comments
every day	Indian #1	Moon	14 hours	+ 14 hours	
every day	Indian #2	Moon	14 hours	+ 14 hours	
every day	Pacific #1	Moon	14 hours	+ 14 hours	Fast
every day	Pacific #2	Moon	14 hours	+ 14 hours	
every day	Atlantic #1	Moon	14 hours	+ 14 hours	
every day	Atlantic #2	Moon	14 hours	+ 14 hours	

missions. The game changing reason for Galactic Harbours is that their strengths will enable a remarkable future for humankind. This recently recognized set of strengths breaks out into two categories that ensure the vision for Space Elevators:

Space Elevator Vision:
The space elevator gives us the road to limitless opportunities while opening up the solar system.

Strengths 1 - <u>Enables Liftoffs towards space</u> with routine, daily, environmentally friendly, inexpensive delivery of cargo and humans:

- Routine [daily launches]
- Permanent infrastructure (no throw-a-ways)
  - Multiple paths when infrastructure matures
  - 24/7/365/50 yrs. [bridge similarities]
  - Massive loads multi-times per week [7 tether climbers per elevator]
  - Cargo segments of 14 metric tonnes each
  - Little impact upon the global environment
  - Does not leave space debris in orbit
  - Safe [no chemical explosions from propulsion]
- Revolutionarily inexpensive to GEO
  - Commercial development similar to bridge building
  - Financial numbers that are infrastructure enabling
  - No consumption of fuel [solar cells will drive the motors for lift]
- Design flexibility for Cargo
  - Opening up design options for space systems
  - No shake-rattle-roll during launch
  - Fewer volumetric restrictions for launch
  - Minimum stressors with slow accelerations
- Environmentally Friendly
  - No burning exhaust with residual hazardous materials into atmosphere
  - No disruption of the ozone layer in the upper atmosphere
  - Improves Earth's environment by accomplishing missions not seen before such as dispersing nuclear waste, sun shades, and moving hazardous operations off-Earth

Strengths 2 - <u>Enables Remarkable Missions</u> only dreamt of before:

- Massive movements to GEO
  - Construction of Large Satellites
  - Recovery and repair of satellites
  - Co-orbiting [floating] at GEO for easy delivery and assembly.
  - Easy delivery to GEO location within a week
- Revolutionary Approach to Interplanetary Destinations - both robotic and human colonization (which require massive support)
  - Massive movement of cargo towards the Moon, Mars and other Solar System Destinations (asteroids, comets, L-5 location, other Moons, planets, etc.)
  - Rapid movement to these locations (as low as 61 days to Mars with release from Apex Anchor)
  - Daily release towards interplanetary missions from Apex Anchor
- New Mission Enabling - Space Elevators can do these in a timely manner!
  - Space Solar Power needs massive spacecraft (5,000,000 metric tonnes) at GEO

o Interplanetary Mission Support (1,000,000 metric tonnes to SpaceX's Colony)
o Moon Village requires massive support with timely delivery required
o Release of Nuclear waste to disposal orbits towards the sun - routinely/safe/inexpensive releases from Apex Anchor
o Sun Shade at Earth-Sun L-1 location for cooling the Earth

The bottom line for Galactic Harbours, and missions within the solar system, is that humanity's hopes and needs to expand beyond the limited resources and environment of our own planet can be realized. Space Elevators are the enabling infrastructure ensuring humanity's growth within and beyond our solar system.

1.6 New Recognition of Demand for Logistics Support:

In the past, the demand for logistics support has not been separated from human deployment to the Moon and Mars. The rocket portal approach to launch support drives the reliability of every vehicle to human ratings. When we are talking about humans to space (to the ISS, the Moon, Mars or a free floating colony) you are talking about a tremendous amount of support cargo delivered to their destinations. The concept of a Moon Village is remarkable and it is encouraging that people are working towards that; but, how much needs to be brought from Earth to house, feed, oxygenate, and entertain humans. One question still to be addressed is what are the demands to support people on the Moon and Mars. If done by rockets, the inefficiencies would dominate the numbers. When Apollo went to the Moon, they put the lunar lander (and ascent vehicle) on the surface which was only 0.5% of total mass of the Saturn V on the pad at liftoff. Now, modern rockets are better at efficiencies and re-usability; however the rocket equation still dominates and can probably not get more than 5% of the total mass on pad at departure towards the Moon. To make it even worse, it takes so many launches and safe landings before we can support people on the surface of the Moon. "NASA is currently planning a series of 37 rocket launches, both robotic and crewed, that will culminate with the 2028 deployment of the first components for a long-term lunar base."[13]

Demand Pull: Chapter Three delves into the project needs to support the diverse missions for Galactic Harbours and Space Elevators and where those needs are headed. During these discussions, three reference missions will be described to layout throughputs for Galactic Harbours to support them. The chapter also lays out several other missions to ensure that nations do not just focus on the single need to support interplanetary missions. This transportation infrastructure will be complementary to rocket portals and will ensure that support cargo is delivered wherever humans decide to go with robotic spacecraft or human ventures.

---

[13] http://www.astronomy.com/news/2019/05/moon-village-humanitys-first-step-toward-a-lunar-colony

Interplanetary Mission Support is an important requirement that will directly drive infrastructure designs. What will enable the robust movement to GEO/Moon/Mars? People will be going beyond LEO, now we must plan to support them. This study should demonstrate that between complementary space access infrastructures, these missions can be achieved. They might be distinctly different infrastructures; but, they are complimentary in their approaches to Interplanetary Support Missions.

> ***Remember:*** This is the transportation story of the 21st century. Reliable, environmentally friendly, routine, safe, and efficient access to space is close at hand. The combination of transportation infrastructures (complementary rockets and Space Elevators strengths) will ensure that missions are supported within global and interplanetary mission needs.

1.7 Conclusions:

During the International Space Development Conference in June of 2019, four Galactic Harbour themes were presented and discussed at length. The themes are descriptive of today's understanding of the Space Elevator's status and are presented here:

- ***Theme One***:   Space Elevators are closer than you think!
- ***Theme Two***:   Galactic Harbour is a part of this global and interplanetary transportation infrastructure
- ***Theme Three***: Space Elevator development has gone beyond a preliminary technology readiness assessment and is ready to enter initial engineering validation testing -- leading to establishment of needed capabilities.
- ***Theme Four***: The magnitude of Galactic Harbour Architecture demands that it be understood and supported internationally.

In addition, the unique characteristics of release from the Apex Anchor leads to the conclusion that interplanetary flight and movement off-planet requires a complementary infrastructure with rockets and Space Elevators sharing missions with each system leveraging their strengths. Rocket portals have tremendous strengths; however, the new concept of a Galactic Harbour Architecture has unique characteristics that ensure they will "enable" interplanetary missions as complementary infrastructure to rockets. These include:

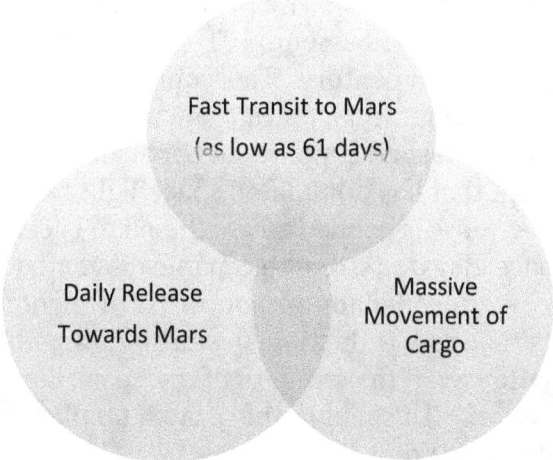

Figure 1.6, Newly Recognized Strengths

The Space Elevator community needs to insert itself into off-planet conversations and express its unique strengths and opportunities. The discussions should now involve movement towards the Moon and Mars. Interplanetary Mission Support is an important future mission for Space Elevators. Galactic Harbours will enable robust mission support off-planet. There are needs to be explained within the concept in engineering, financial and programmatic terms. Some conclusions from these analyses are stimulating!

- Galactic Harbours will enable robust missions to Moon and Mars!
- Only Space Elevators can deliver the requirements of logistics equipment and supplies to the Moon and Mars
- Colonization on Mars cannot happen without the logistics support of Space Elevators!
- Launch Windows to Mars every 26 months - be Damned! [a favorite]

Reliable, routine, safe, environmentally friendly, and efficient access to space is close at hand. The Space Elevator is the Galactic Harbour, and an essential part of a global and interplanetary transportation infrastructure. The requirements (needs) that are being discussed today can be fulfilled with a Galactic Harbour infrastructure enabling the movement beyond Earth throughout this new century. One realization that has not been recognized in future planning is that mission approaches have new alternatives. One option does not require building everything on-site. In-situ resources are necessary; but with Galactic Harbours, the delivery of cargo, supplies, habitats, water, air and all future needs can continue to be procured from the Earth through routine, inexpensive, reliable and daily deliveries. This transportation story of the 21st century changes the previous assumptions related to Off-Planet Movement! Delivery of mass to off-planet destinations from Earth is no longer onerous - but "easy, can-do."

Operational collaboration between the Galactic Harbour's Space Elevators and a variety of space launch competencies is essential. Together they will become the space transportation solution of this century. This combination of strengths enables movement of massive amounts of material for our journeys throughout the Solar System. The ability to produce needed supplies on Earth and ship them in a timely manner - changes the concept that has dominated spaceflight from the beginning. The new thought process can be "mass is good." (an anathema to rocket launching) The complementary rocket/space elevator strengths enable movement of massive amounts of material to space as needed for our journeys throughout the Solar System. This report shows that now is the time to start building the right combination of strengths! However, the strength of the Space Elevator concept is that it can "beat the rocket equation." This makes the growth off planet reasonable. Figure 1.7 (next page) makes this point.

This study report illustrates many new conclusions relative to Space Elevator and Galactic Harbour strengths and drives a new assumption for future off-planet movement.

"A Network of Galactic Harbours enables humankind's movement off of Planet Earth"

Can the Rocket Equation be "beaten?" - Only by the Space Elevator!

The bottom line - space elevators are compatible and complementary to rocket architectures. The future needs both communities to work together. However, the first step is to help the rocket community understand the strengths of space elevators - "we can beat the rocket equation." Wikipedia defines the rocket equation as:

> "The mathematical equation that describes the motion of vehicles that follow the basic principle of a rocket: a device that can apply acceleration to itself using thrust by ***expelling part of its mass with high velocity*** can thereby move due to the conservation of momentum."

To compare the delivery of payloads to LEO and GEO, information is gathered from the wiki. (information 26 May 2020). The comparisons of rocket equation results shows that delivery to LEO is less than 5% of mass at pad and delivery to GEO (or trans-lunar insertion) is about half that while delivery to GEO or Apex Anchor by space elevator is 100% of payload at Earth Port. There are a few items that need to be recognized when looking at the rocket equation:

- there are no "cost factors" inside the rocket equation
- there are no reusability factors in the rocket equation

As a result, the rocket community can decrease the cost and leverage reusability of rocket stages to increase operational efficiency. However, those actions do not "improve the rocket equation." The Tsiolkovsky rocket equation still responds to that critical factor called gravity. The Earth's gravity numbers have a consistent impact on efficiency at liftoff and flight - draconian.   Another comparison would be with reference to the Moon. The Saturn 5 deposited 0.5% of pad mass on to the surface of the Moon and returned 0.18% to the ocean upon completion of the mission. Those are tough numbers to build around. However, if you raise 20 MT to GEO and then Apex Anchor with electricity, you have beaten gravity $(1/r^2)$ and added tremendous velocity (7.76 km/sec) for release towards mission destinations.

Yes, the Space Elevator "beats" the Rocket Equation.

Figure 1.7, Rocket Equation Beaten (expanded in Appendix A)

1.8 Chapter Breakout: This study report is broken down as follows:

Chapter 1   Introduction - Enabling Interplanetary Missions: This chapter established some significant findings: (1) Change of Vision for Interplanetary Movement when delivery of mass is inexpensive, timely and supportive. This new vision of Galactic Harbour architectures will change the "thinking" for off-planet migration - We can bring it with us!, (2) Movement off-planet will require complementary capabilities, such as rockets portals and Galactic Harbour infrastructures, each with their own strengths and short-falls. and, (3) Discussion of mission needs when analyzing mass needs at various locations. This discussion will start the analysis of "how much carrying capability" is required by each supportive infrastructure; when, to where, and their priority. In addition this chapter initiated the discussions around newly recognized Galactic Harbour strengths for Interplanetary Mission Support - Fast Transit, Massive Movement of cargo, and Daily releases towards solar system locations.

Chapter 2   Visionary Look at Galactic Harbours and Interplanetary Missions: This chapter establishes a vision to support the needs to move off-planet with Galactic Harbours. This 2020 Vision is a portrayal of the fulfilled transportation story of the 21st Century. It is the extension of recent experiences and the manifestation of humankind's initial expansion into the rest of the Universe. This is an unabashed explanation of what can be seen with 2020 foresight.

Chapter 3   Enhanced Mission Needs for Interplanetary: This chapter introduces the mission needs of the future. The discussions start with current missions (mostly in Earth's orbits) and expands towards the future of interplanetary mission support. It focuses upon the development of the reference missions as a method to demonstrate the future demand pull. These are: Space Solar Power from GEO, Moon Village, and SpaceX's Colony on Mars.

Chapter 4   Arizona State University Research Approach and Results: This chapter discusses the remarkable research conducted at ASU with Dr. Matthew Peet as lead with over ten students participating. They focused upon the following questions: 1) "How often can we expect to deliver payloads to Mars from Space Elevator's Apex Anchors?" What are the launch windows? - weekly or daily?   2) "What are the transit times when departing the Earth's sphere of influence with three times the energy of past chemical rockets? and, 3) "With those two discussions initiated, the concept of multiple metric tons of mission support logistics arriving at Mars (or the Moon) per day is remarkable. How will that change the approach to support human movement off-planet? Their overall conclusion was: Interplanetary travel is greatly enhanced by release from the Apex Anchor in time of flight, required delta velocity and launch opportunities.

Chapter 5   Projected Throughput for future Galactic Harbours: This chapter focused upon the capacity of Galactic Harbours. It starts with a single tether Space Elevator initial capability and grows the network of Galactic Harbours to include six Space Elevators, within three Galactic Harbours, with mature capabilities. The projected throughput establishes the need for Space Elevators and Galactic Harbours to be included in the discussions about the future.

Chapter 6   Logistics Supply Chain - Galactic: This one places the Galactic Harbour within the global logistics supply chain. The new movement off-planet will require that the global transportation infrastructure embrace this new capability to extend beyond terrestrial routes to include the solar system three-dimensional arena.

Chapter 7   Conclusions and Recommendations: Chapter seven lays out the study report's conclusions and then discusses recommendations for the future.

In addition, a List of References and several Appendices:A - Defeating the Rocket Equation, B - Encyclical Vision, C - SE ready to proceed Questions? D - ISEC Study List, E - IAA Study List, F - What is ISEC, and G - Lexicon of Terms.

Chapter 2 - 2020 Vision[14] - Where Will Galactic Harbours Be?

2020 Vision:

The Space Elevator story is still being written.
The Apex is where the Galactic Harbour meets the Shoreline of
Outer Space and Where the "Transportation Story of the 21st
Century" meets the "Final Frontier."

2.1 Introduction:

This report projects what the Space Elevator and Galactic Harbour will become. This 2020 Vision is a portrayal of the fulfilled transportation story of the 21st Century. It is the extension of our experience and the manifestation of humankind's initial expansion into the rest of the Universe. This vision is an explanation of what is projected with 2020 foresight. The current architectural team sees a network of space elevators as a visionary megaproject that will have mega impacts.

In the recent past, the Space Elevator community has taken this simple concept and defined a rather capable Space Elevator based Galactic Harbour. This approach has always been one called "Technology Push" and has usually led to technological developments responding to near term needs. Now that we have established the baseline Space Elevator with the recognition that we are moving out of the technological feasibility question, the need is to look towards the future and identify the major missions to be fulfilled by Galactic Harbour transportation infrastructures. This approach is called "Mission Pull" (or "Demand Pull") and will enable us to look towards the future and where the concept is going. The point of this discussion is beyond the idea that Space Elevators are Tech Ready; beyond the reality of engineering requirements; and, even beyond the recent realization that Galactic Harbours and their Space Elevators are the essential and enabling parts of a logistics support space infrastructure. Thus, this study will look beyond Tech Ready towards what the mature Space Elevator can become. It isn't the baseline architecture. It is the authors' projection of our baseline all grown up. The current concepts will grow and enlarge their capabilities and engineering performances to support projected future

---

[14] Fitzgerald, M and V. Hall Portions of this chapter leveraged vision statement, Dec 2019 (see appendix).

missions, or in system engineering terms, system maturity. It will be the transportation story of the 21st Century.

2.2 Background:

By the middle of the 21st Century, *Galactic Harbours* will provide a nexus for connecting the two dimensional transportation systems of rail, highway, shipping, and air to the third dimension -- space. This three dimensional infrastructure will provide a safe, reliable and cost effective means to move material and people into orbit and beyond -- from any place on Earth through this networked transportation infrastructure. Appendix B, "*Encyclical*," expands the future vision for a mature set of Galactic Harbours around the Earth's equator satisfying many customers with varied missions, such as: Mission to Mars and the Moon, Space Based Solar Power at GEO, and many more. This complexity of future missions leads to worldwide networks of Galactic Harbours delivering cargo from the Earth to multiple destinations out in space.

2.3 Earth's Galactic Harbours:

The future sees a network of diverse elevators, spread around the Earth; however, they differ by more than location. Their respective engineering differs by their purpose. Cargo trans-shipment and vertical cargo transport are common to each, but the Harbours are optimized for their respective specific purposes. In addition, cargo transfer will be supporting numerous commercial developments up and down the Space Elevators. Some of these diverse missions are:
   o GEO Orbit factory support: Space based solar power, satellite servicing – refuel, repair, upgrade, assembly of large spacecraft and space facilities, and other on-orbit enterprises.
   o Interplanetary travel support: Flight systems final assembly, departure mission readiness and checkout, departure and arrival operations, fueling of special thrusters for long distance operations, massive cargo and material handling operations.
   o Research and science support: sensing systems, flight operations within the elevator regions, computers and communications, testing and test data collection facilities, robotic flights to all areas in our solar systems.
   o Tourism support: human rated activity, elevator regional flight operations, safety and comfort.
   o Government support: space and region debris mitigation, space traffic management and control, law enforcement, first responders, safety assurance operations, sovereign issues and relations.

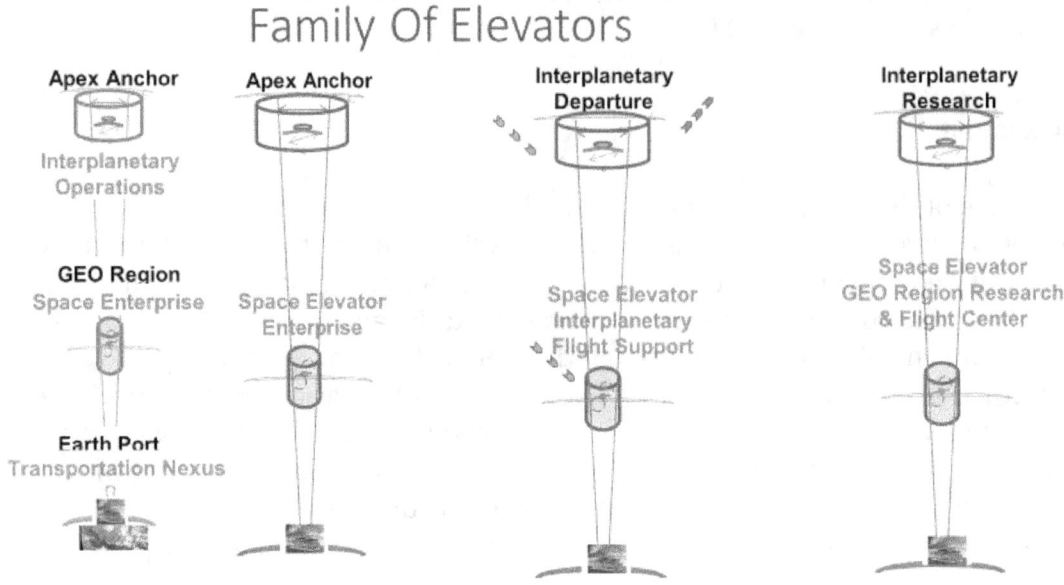

Figure 2.1, Family of Space Elevators

Some of these elevators are stand alone, and some are hybrid elevators sharing operational regions with one another. To some extent, governmental use is embedded in all elevators. Travel from one elevator to another is on enhanced pathways along and across the geosynchronous belt. Trade between the elevators is robust - especially for delivery of key repair items -- FedEx, UPS, Uber and Lift aloft. Some of the elevators are human rated and some are purely robotic. Some are small and speedy, and some have immense throughput. Indeed, the two elevators inside each Galactic Harbour do not have to support the same mission. This is called the "Family of Space Elevators," as they support so many diverse missions. The purposes of Earth's Space Elevators are broken down into three basic ideas: 1) deliver cargo to the enterprises assembling along the geosynchronous belt near the Space Elevator's GEO Regions; 2) deliver satellites to Earth orbits supporting various missions; and 3) support interplanetary flights from the APEX Regions to the Moon, Mars, and elsewhere.

2.4 Vision for Transportation Infrastructure for Interplanetary Mission Support:

The authors' vision of Earth Based Space Elevators is a series of Galactic Harbours along the equator supporting various missions as a three dimensional transportation infrastructure. Within that series of Galactic Harbours will be support to interplanetary missions to the Moon and Mars, asteroids, other planets, and to L-5 type human colonies. The complementary aspects of rocket propulsion and space elevator releases will enable the human race to move towards their desired destinations with confidence that they will have all the support they need. The magnitude of this portrayal is humbling. This future is marked by the needs of humanity and the natural instinct to explore and expand our horizons. To execute this sort of remarkable future, the authors have developed an idea leading to a vision of a mature transportation infrastructure around the tipping point of the 21st century. Our vision of the Galactic Harbour is at the horizon, our descriptions of Galactic Harbour operations are beyond that. The team sees something else beyond the idea that the Space Elevator is Tech Ready and beyond building the first Elevator and then the first Galactic Harbour. The team realizes that Galactic Harbours, and their Space Elevators, are the essential and enabling part of a logistics support space infrastructure. This report is a projection of the network of Galactic Harbors all grown up. The Space Elevator concept is Tech Ready. What is presented here is a portrayal of how the concept will mature to support the interplanetary travel objectives of this century and the next.

As the concept of a Galactic Harbour progressed, the recognition of "Demand Pull" created a series of future architectures responding to needs and capabilities. Across this report the following time periods will be used as explanations for the growth of this transportation infrastructure and solar system expansion. This robust vision of the future is responding to the realization that Space Elevators can be developed and that the demand pull for movement off-planet will be well beyond even the capabilities of mature Galactic Harbours.

- Near Future - a Master Galactic Harbour: The initial set of two Space Elevators will show remarkable strengths. This master Galactic Harbour will be supporting Initial Operational Capability (IOC - 14 tonnes per day) on each Space Elevator while developing business enterprises across the cosmos. These activities can be "seen" as occurring in the near future (2035 to 2040).

- Near the Horizon - a network of Galactic Harbours: The competitive growth of these transportation infrastructures will expand to at least three Galactic Harbours around the equator with IOC capable Space Elevators. This is shown in the image below.

- Just over the Horizon - a Network of FOC Galactic Harbours: This mature Galactic Harbour transportation infrastructure will consist of three (or more) Galactic Harbours distributed around the equator with two space elevators within each. The capability to transport 79 metric tonnes per tether climber per day will have matured during the developmental phases. As shown in this report, this remarkable capability will still not satisfy all the needs to move off-planet and support Earth's other orbital missions. The possibilities for further growth are endless.

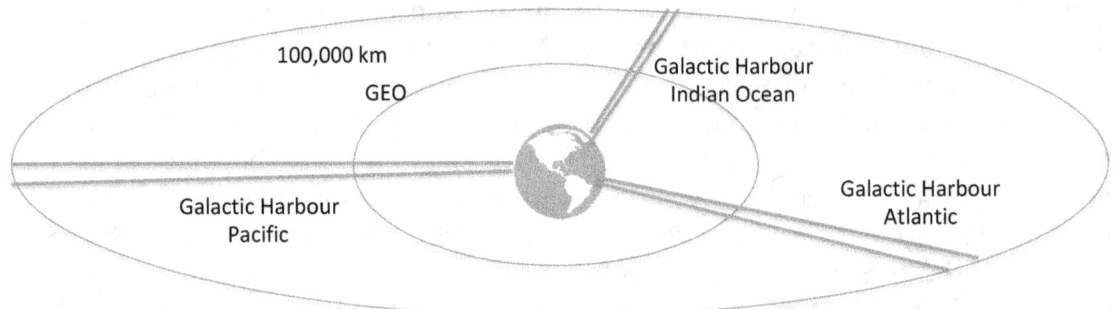

Figure 2.2, Three Galactic Harbours - A Vision At the Horizon

2.5 The Solar System's Elevators:

As interplanetary flights from the Apex Anchor will use latent velocity, (derived from Earth's rotation - transformed into radial speed at departure from 100,000 km altitude) for its Interplanetary Mission Support, the question is who will receive the massive cargos being sent. It seems obvious that there should be Space Elevators established around the Solar System as part of Galactic Harbour transportation networks. Elevators should operate attached to the Moon, near Mars, and on key asteroids within the asteroid belt and elsewhere. The authors see immense cargo craft moving from elevator to elevator delivering supplies and equipment, and returning with raw materials for processing in one of several GEO regions and later to Earth. This is the third dimension of trade, commerce, transportation, and humankind. This aspect of our future vision will be saved for a separate ISEC study report in the future.

2.6 Conclusions:

Our 2020 Vision is a portrayal of the fulfilled transportation story of the 21st Century. It is the extension of our experience and the manifestation of humankind's initial expansion into the rest of the Universe. This is an unabashed explanation of what we see with 2020 foresight. The magnitude of this portrayal is humbling. It will be accomplished in the time to come. It is a well-marked destination - marked by the needs of humanity. It was seen before; yes, seen by those who preceded us - and it will be built by those who follow.

The Galactic Harbour Network is Earth's lifeline to the future

Chapter 3 - GEO and Interplanetary Missions Needs

3.0 Complementary Architecture:

The operational collaboration between the Galactic Harbour's Space Elevators and a variety of rocket launch competencies is essential. Together they will compose the space transportation architecture for this century. This future architecture enables movement of massive amounts of material for our journeys throughout the Solar System as well as insuring movement of people rapidly through radiation belts. One obvious change in planning with this complementary architecture of both rockets portals and Space Elevator infrastructures is you don't need to depend heavily on the local resources at your destination to survive. With Galactic Harbours providing massive cargo movement, the restrictions of rocket deliveries go away and the traveler can depend on support from Mother Earth facilities.

The restriction of rockets, only delivering 3% of launch pad mass to interplanetary destination, turns into a strength of Galactic Harbour Earth Ports as 100% of lift-off cargo is released towards its destination.[15] This relates to 70% of liftoff mass.

3.1 Introduction:

The future reality seems to be that when humanity decides to move massive amounts of equipment to GEO, the Moon, and then Mars, there will be a tremendous need for logistics support [mission support equipment, habitats, food, oxygen, fuel, and transportation modes]. As the Space Elevator will be routine - but not fast there will be a great need for transporting people by rocket [especially routinely and rapidly through radiation belts]. The expansion of missions at GEO will only increase the needs for critical mission support from across the globe. The realization that we will be placing boots on the Moon in the near future assumes that there will be plenty of supplies to support those travelers. Mission support to the lunar surface will be intense and will require a massive capability. And, of course, the mission to place people on Mars and to start a colony is now becoming real. Massive amounts of supplies and equipment must be on the surface prior to the first human arrivals. The envisioned Missions to Mars and Moon are of incredible size and scope. In addition, it is becoming obvious that the growth at GEO will be remarkable, especially when one includes the need to supply the Earth with electrical power in an efficient and environmentally friendly manner - such as Space Solar Power.

---

[15] Breakout roughly 70% payload (14 tonnes cargo) with 30% tether climber (6 tonnes)., or 70% of liftoff weight.

3.1.1 Diverse Missions: The Galactic Harbours around the equator during the middle of this century will support traditional missions, new missions to traditional orbits and the remarkable expanding missions to the Moon and interplanetary destinations. During this chapter, the listing of several missions will be shown while focusing upon three reference destinations that will help "size" the capacity needs for mission support. Massive movement, of cargo and support infrastructure to locations in GEO and beyond, is a strength of Space Elevators. The enterprise of Galactic Harbours and support to individual missions will ensure aggressive growth of satellites to GEO and beyond. The general breakout of missions would include (but not be limited to):

- Traditional Geosynchronous Orbit Missions: Traditional Satellites such as weather, communications, and governmental missions will be enhanced and expanded as access becomes easier and cheaper. There are over 400 active GEO satellites as of October 2018 conducting these missions.[16] When the cost and simplicity of operations goes way down, this number will escalate.
- Revolutionary Geosynchronous Orbit Missions: New missions will be supported such as refueling and repair of ailing satellites, construction of new systems larger than a single payload from tether climber or rocket fairings, and new enterprises not even thought of during the first three decades of this century. This will be a huge growth area when people realize the opportunities. One of these missions will be Space Solar Power transmitting electrical power to the surface of the Earth at remarkable low prices. (this will be one of the three reference missions developed inside this study.)
- Lunar and Interplanetary Missions: These include two reference missions (Moon Village and Mars Colony; see below for expansion); robotic missions to anywhere in our solar system; exploration missions to Moon and Mars and beyond; human missions to L-5 type colonies; and robotic missions beyond solar system - on to the stars will be discussed.

As shown above, the Galactic Harbour is a major portion of what is now being called the Mosaic of Space. This transportation infrastructure will become a key element of the newly forming strategic mosaic of space. Mankind is no longer just going to space, they are conducting exploration, research, military operations, trade and commercial enterprises. The placement of Galactic Harbours inside the strategic mosaic of space will ensure that the exploitation of this tremendous new access to space will leverage the lessons of history and enhance the safety of the enterprise. The strategic mosaic of space is taking form today.   It is composed of trade, enterprise, research, exploration, and military protection. The ability of the Space Elevator transportation infrastructure to be a logistics giant will ensure that this movement off-planet will result in an economic engine on (and near) Earth. The codification of the engineering transportation infrastructure will solidify the segment-to-segment relationships and

---

[16] Wikipedia, 23 March 2020. https://www.google.com/search?

support the satisfaction of system level requirements in preparation of design activities.

In 2014, the International Academy of Astronautics published a study report entitled, "Space Elevators: An Assessment of the Technological Feasibility and the Way Forward."[17] The next chart shows the diversity and demands from future missions, as seen by those 47 academicians. This six-year old listing has some intriguing missions and their demands for Space Elevator delivery of cargo to various destinations. It is a starting point for discussions about loading. The numbers show the movement of metric tonnes per year for each projected mission.

Table 3.1, Projected Demand (metric tonnes per year)

Demand in Metric Tons	2031	2035	2040	2045
Space Solar Power	40,000	70,000	100,000	130,000
Nuclear Materials Disposal	12,000	18,000	24,000	30,000
Asteroid Mining	1,000	2,000	3,000	5,000
Interplanetary Flights	100	200	300	350
Innovative Missions to GEO	347	365	389	400
Colonization of Solar System	50	200	1,000	5,000
Marketing & Advertising	15	30	50	100
Sun Shades at L-1	5,000	10,000	5,000	3,000
Current GEO satellites + LEOs	347	365	389	400
Total Metric Tons per Year	58,859	101,160	134,128	174,250

3.1.2 Demand for Logistics Support: The demand for logistics support had not been separated from human deployment to the Moon and Mars. This single approach (rockets only) in launch support drives the reliability of every vehicle to human rating. When discussing humans to space (to the ISS, to the Moon, to Mars or a free floating colony) the reader must recognize the need for tremendous amount of support cargo delivered to their destinations. The concept of a Moon Village is remarkable and it is encouraging that people are working towards that. An important question that must be answered early in the planning process is how much do we have to bring from Earth to house, feed, oxygenate, and entertain humans. One question still to be addressed is what are the demands to support people on the Moon and Mars. If done by rockets, the inefficiencies would dominate the numbers. When Apollo went to the Moon, they landed the lunar lander (and ascent vehicle) on the surface as only 0.5% of total mass of the Saturn V on the pad at liftoff. Today, modern

---

[17] Swan, P., Raitt, Swan, Penny, Knapman. International Academy of Astronautics Study Report, Space Elevators: An Assessment of the Technological Feasibility and the Way Forward, Virginia Edition Publishing Company, Science Deck (2013) ISBN-13: 978-2917761311

rockets are better at efficiencies and re-usability; however the rocket equation still dominates and the answer can probably not get better than 5% of total mass on pad at departure towards the Moon's surface. The newest proposal, Starship, will have 100 Metric tonnes of payload when it lifts off from its launch pad at a total mass of 5,000 Metric tonnes[18] (or 2% of payload to launch mass), although it is totally reusable. To make it even worse, it takes a tremendous number of launches, fuel transfers in LEO, and safe landings before we can support people on the surface of the Moon.

> "NASA is currently planning a series of 37 rocket launches, both robotic and crewed, that will culminate with the 2028 deployment of the first components for a long-term lunar base."[19]

3.1.3 Mission Needs: If one were to take the hugely successful cruise ship analog, one could say that it takes 12 metric tonnes per person to support them for one week. The number comes from a massive vessel (100,000 tonnes weight) with lots of humans (8,500 guests and crew). Not sure this is a good parallel; but, it illustrates the tremendous needs of humans to live their daily lives. The real question is how much support does an individual need when going to Mars or the Moon, on-average? The demand should be spread out to areas such as - preliminary exploration, upgrade capabilities to include scientists-engineers-builders in longer stays, and then continuous demands to maintain the colony as a full time environment. These numbers need to be quantified to enable realistic planning. As colonies form, the need for support will increase at a tremendous rate. The mission support mass per day required to be delivered to the Moon, Mars and other destinations will over stress any rocket only architecture as well as being exorbitantly expensive. Space Elevators allow mission support growth to accelerate with three major improvements:

- Inexpensive and routine (daily) massive movement of mission support equipment
- Tremendous opening up of launch windows (daily to weekly towards Mars)
- While reducing travel times (fast transits as short as 61 days to Mars)

3.2 Reference Missions:

Specific needs for individual missions can be assessed with some estimates already expressed. Over the last few months several "need" statements have been expressed to show the magnitude of the cargo required to establish appropriate missions. This study has looked at three cases as a method of illustrating the magnitude of the effort to fulfill the needs of humans in the next few decades. Each of these is described quickly below as starting points for the discussion of Mission NEEDS. Two of the

---

[18] wiki, 23 March 2020, https://www.google.com/search?client=safari&rls=en&q=starship&ie=UTF-8&oe=UTF-8

[19] http://www.astronomy.com/news/2019/05/moon-village-humanitys-first-step-toward-a-lunar-colony

three have already established a baseline need for support in the sense of mass to destination. These will be discussed which will result in a remarkable understanding of demand for Space Elevator infrastructure capacity.

Reference Missions: To place this whole study in perspective, the comparison lays out three Reference Missions:
- GEO Base - Space Solar Power (IAA study + Mankins' study activities)
- Moon Base - Lunar Village (ESA ++)
- Mars Base - SpaceX Colony

By starting with known activities, the estimates of logistics demands should be easier to quantify. The refinement of the Reference Destination Missions will be exhaustive showing what initially is needed by the first off-planet community and then required for its continued growth. Further explanation of the three mission destinations are as follows:

3.2.1 Space Solar Power:

"Space solar power can solve our energy and greenhouse gas emissions problems. Not just help, not just take a step in the right direction, but solve."[20]

This mission of a geosynchronous orbit destination has the potential to solve most of the Earth's energy problems in a truly environmentally friendly manner: providing electrical energy without burning fossil fuels. The following few paragraphs are taken from a book by John Mankins outlining the engineering feasibility of the design as well as the need for multiple space systems supplying up to 10% of the needed electrical power required by humans around 2060 - globally, and inexpensively. The system was described in his book as:

"At an altitude of 22,240 miles above the Earth (35,786 km), a great platform orbits: collecting with vast, mirrored wings the continuous torrent of sunlight always available in space, redirecting and focusing that energy onto concentrating photovoltaic solar arrays that convert it into electrical power. With great efficiency, this tremendous platform uses the power of the Sun to generate a coherent stream of radio energy, and then wirelessly transmits that power with minimal losses down to highly efficient receivers the size of airports on the ground. The platform high in space gathers in over 5,000 megawatts of sunlight, and delivers – day and night – over 2,000 megawatts of clean, near zero-carbon electrical power to customers as needed, anywhere within an area the size of a continent or more." [21]

---

[20] Mankins, John, The Case for Space Solar Power, Virginia Edition Publishing Co. Dec 2013.

[21] Mankins, John, The Case for Space Solar Power, Virginia Edition Publishing Co. Dec 2013.

The Promise:[22]   "Reliable and affordable energy is fundamental to our global society. It is only through the availability of vast amounts of cheap energy – largely from fossil fuels – that the world's population grew from less than one billion in 1800 to more than seven billion in 2013. Most of that energy is used in the form of electricity, with most of that electricity generated through the burning of coal. As the world's economy has grown over the past three decades – raising the quality of life in China, India, and many other countries – the need has become all the more urgent to increase the energy available for industry, transportation, for heating and cooling, for personal use, and so on. In addition, during the past several decades concerns have emerged that society's overwhelming dependence on fossil fuels is driving other kinds of change: growth in atmospheric concentrations of greenhouse gases (such as carbon dioxide and methane) and increases in average global temperatures. If left unchecked, most scientists now believe that increasing concentrations of greenhouse gases in the atmosphere will alter the global climate by the end of this century, with sweeping impacts on societies across the globe. As a result, there is great interest in finding new, more sustainable alternatives to fossil fuel- based energy supplies. Of the non-fossil fuel alternatives that exist – hydroelectric power, nuclear power, wind power, wave energy and others – one of the most accessible and intuitively attractive is solar power. However, despite significant advances in performance, reductions in cost, and dramatic growth in the total deployed capacity in recent years, ground-based solar power remains largely a niche technology, providing only a small portion of society's energy needs. Fortunately, there is a promising alternative to conventional ground-based solar power systems, albeit one that is a relatively unknown: Space Solar Power.[23] The energy in the sunlight found in space near Earth is considerably greater than that which remains in sunshine after it passes through the atmosphere, even on a clear day. In fact, the power intensity of sunlight in space is about 1,368 watts per square meter, as compared to only about 1,000 watts/m2 at noon on a clear day near the equator – a drop of about twenty-seven percent (27%). This initial attenuation of sunlight is compounded by additional factors: the day-night cycle, a reduction averaging roughly sixty percent (60%); changes in the available sunlight due to the weather, a reduction of twenty percent (20%) for light clouds, but up to seventy-to-eighty percent (70%-80%) for heavy clouds; and changes due to the seasons, of up to sixty percent (60%) or more, depending on the latitude of the site. The combination of these factors results in the available solar energy in space at around GEO or above being about ten times greater than the best average available at most locations on Earth. Looking at the data, it is clear that the available solar energy at a typical location on Earth on the worst day – even in the best month – is a tiny fraction of the solar energy available in space nearby."

---

[22] Mankins, John C., "The Case for Space Solar Power," Virginia Edition Publishing, 2014.

[23] Space Solar Power (SSP) is the generic term for this technology that has been used for the longest time. It refers to either space applications or systems that deliver energy to markets on Earth. Another term sometimes used for an SSP system that delivers energy to Earth is "Space-based Solar Power" (aka, "SBSP"), which was used by the 2007 assessment performed by the US Department of Defense (DoD) National Security Space Office (NSSO).

One conclusion from Mr. Mankins' book is:

"It is crucial that the systems used for space transportation must be transformed in order for space solar power to become economically viable."
"It is crucial for the world to identify, develop, and deploy affordable and sustainable new energy sources. This need is driven by a number of factors, including three critical ones: (1) demand for energy to enable economic growth for a still-increasing global population, (2) concerns regarding the long-term accumulation in Earth's atmosphere of fossil fuel-derived greenhouse gases, and (3) the prospect that during the coming decades annual production of petroleum (and possibly other fossil fuels) will peak and begin to decline."

Figure 3.1, Space Solar Power System Design[24]

This ISEC study must look at the massive payloads needed to be delivered to the GEO belt if the mission is endorsed. Space Solar Power deserves to be evaluated as a mission that would require massive space mission support, while helping to save the environment. The impact on humanity for this mission is critical and should be included in the sizing of future mission support infrastructure. During the 2019 International Astronautical Congress in Washington DC, ISEC met with Dr. Mankins and discussed exactly what he would need to succeed in his mission to implement the Space Solar Power architecture he envisions. His answer was 5,000,000 metric tonnes as the required mass to GEO to enable support to 12% of

---

[24] Mankins, John, The Case for Space Solar Power, Virginia Edition Publishing Co. Dec 2013.

humans by 2060.[25] This requirement seems to be a large quantity; however, if the human condition is to improve because of Galactic Harbours, this mission must be supported.

Table 3.2, Reference Destination - Geosynchronous Orbit

Reference Mission	Metric Tonnes to Destination	Comment
Space Solar Power	5,000,000	Power to 12% of Earth's population in an environmentally friendly manner.

3.2.1   SpaceX Colony on Mars: Recently, Elon Musk was discussing his plan to go to Mars and create a colony.

"By talking about the SpaceX Mars architecture, I want to make Mars seem possible—make it seem as though it is something that we can do in our lifetime. There really is a way that anyone could go if they wanted to. I think there are really two fundamental paths. History is going to bifurcate along two directions. One path is we stay on Earth forever, and then there will be some eventual extinction event. I do not have an immediate doomsday prophecy, but eventually, history suggests, there will be some doomsday event.

Figure 3.2 SpaceX's Colony[26]

The alternative is to become a space faring civilization and a multi-planetary species, which I hope you would agree is the right way to go. So how do we figure out how to take you to Mars and create a self-sustaining city—a city that is not merely an outpost but which can become a planet in its own right, allowing us to become a truly multi-planetary species? Sometimes people wonder, 'Well, what about other places in the solar system? Why Mars?' It really only leaves us with one option if we want to become a multi-planetary civilization, and that is Mars. We could conceivably go to our moon, and I actually have nothing against going to the moon, but I think it is challenging to become multi-planetary on the moon because it is much smaller than a planet. It does not have any atmosphere. It is not as

---

[25] Mankins, John, personal conversation with P. Swan, at IAC, Washington.D.C. Oct 2019.

[26] Musk, Elon, "Making Humans a multi-Planetary Species," New Space, Vol 5, No 2.

resource-rich as Mars. It has got a 28-day day, whereas the Mars day is 24.5 hours. In general, Mars is far better-suited ultimately to scale up to be a self-sustaining civilization."[27]

Supporting a 100,000 person colony on Mars is the vision of SpaceX and plans are being formulated now (Recently Mr. Musk has been discussing a million persons Colony on Mars).   Mr. Musk stated that he needs 1,000,000 Metric tonnes of support for his Colony.[28] The question is how does one deliver one million tons of anything to the surface of Mars? How does the 26-month launch window effect the delivery of assets? Another interesting aspect of the movement to Mars by Elon Musk is that he plans on his StarShip having a capacity to move 100 people (100 to 150 is sometimes quoted) at a time towards Mars with a capability for an extra 100 Metric tonnes (110 to 150 sometimes quoted) of support equipment. That turns out to be only one metric tonnes of supplies for each person for the next two years (remember they are still on 26 month between launch window restrictions with rockets). Of course the explorers will be smaller groups with larger portions of supplies and housing. If we go back to Elon Musk's statement of 1,000,000 metric tonnes of supplies for his colony, then we will need to provide a massive movement of support equipment from Earth to Mars. With Mr. Musk's calculation, the extra supplies for each StarShip would take 10,000 trips. Mr. Musk recently expanded upon his plan to include: 1,000 Starships, three launches each day across the 26 month launch window, 100 tonnes and 100 people per Starship towards Mars, stored in Low Earth Orbit until launch window opens up, and then ship all 1,000 starships towards Mars inside the 30 day window. [29]

In comparison, if we used the initial operational capability of six Space Elevators (three Galactic Harbours), the delivery of 1,000,000 Metric tonnes would take 34 years. However, as our infrastructure will grow in capability, the real estimate using the full capability of future Space Elevators, would be more like five years to supply the full million tons. Permanent transportation infrastructure is the answer, if the need is to support a colony on Mars or even the Moon.

Table 3.3, Reference Destination - Mars & GEO

Reference Mission	Metric Tonnes to Destination	Comment
Space Solar Power	5,000,000	Power to 12% of Earth's population in an environmentally friendly manner.
Mars Colony	1,000,000	Supporting a Colony with logistics has been underappreciated in the movement off-planet

---

[27] Musk, Elon, "Making Humans a multi-Planetary Species," New Space, Vol 5, No 2.

[28] Musk, Elon., Quotation from CBS's Sunday Morning Show, 21 July 2019.

[29] McFall-Johnsen, M, Dave Mosher, "Starship rockets every day and creating 'a lot of jobs' on the red planet," Jan 17, 2020 Business Insider.

3.2.3 Moon Village: The Moon Village is an initiative at the international level that is looking at permanent presence on the Moon.[30] "The project, dubbed Moon Village, first made headlines back in 2015, when the newly appointed Director General of the ESA, Johannn-Dietrich Woerner, told the BBC about his vision. 'Moon Village shouldn't just mean some houses, a church, and a town hall,' he said at the time. 'This Moon Village should mean partners from all over the world contributing to this community with robotic and astronaut missions and support communication satellites.' Though Woerner's comments gave the Moon Village concept its initial momentum, since then, much of the cooperation and planning for the lunar outpost has been spearheaded by the Moon Village Association (MVA),, a non-governmental organization with about 150 members spread across nearly three dozen countries. Working closely with the ESA, the purpose of MVA is to foster collaboration between nations and organizations spread all over the world — especially in developing countries — to help bring visions of a permanent lunar base to fruition. Moon Village is not just going to be another International Space Station on the Moon, said Senior Space Policy Adviser and President of the Moon Village Association, Giuseppe Reibaldi, in a presentation. Instead, Moon Village aims to be a collaboratively designed and expandable permanent lunar settlement that can serve as: an innovative research hub for industry and academia alike; an otherworldly destination for commercial travelers; and a proving ground for future crewed bases on Mars and beyond. 'The goal of the organization is to go beyond space, to go beyond the stars,' said Reibaldi, 'because this is a step for all humanity, and nobody needs to be left out.' Currently, stakeholders plan to build Moon Village near the lunar south pole. In particular, they want to target somewhere close to the rim of Shackleton Crater, which boasts areas that are almost perpetually bathed in sunlight -- ideal because it provides an ample source of energy via the tried-and-true technology of solar panels. In addition to copious amounts of available sunlight, at about 2.5 miles (4 kilometers) deep, the floor of Shackleton Crater is constantly cloaked in shadow. This, means it's the perfect place to hide valuable, sunlight-sensitive volatile resources like water-ice. Such a find nearby could provide residents of Moon Village with drinking water, breathable air, and a potential source of rocket fuel, as long as we have the means to unlock it. According to the plan, Moon Village will be built in incremental steps.

---

30 Moon Village: Humanity's first step toward a lunar colony?
 https://astronomy.com/news/2019/05/moon-village-humanitys-first-step-toward-a-lunar-colony

Figure 3.5, Shackleton Crater (NASA image)

First, individual landers will descend to the lunar surface near Shackleton Crater, at which point they'll each deploy inflatable modules. These modules, which are expected to be up to four-stories tall, would eventually serve as workspaces, residential areas, scientific labs, industrial sites, and environmental habitats. The inflatable modules, however, cannot effectively shield humans from harmful radiation, temperature swings, or micrometeorite strikes. So, the ESA and MVA are currently working with private industries to design robots that can 3D print a protective shell around each structure using readily available regolith taken from the lunar surface. Europe, NASA, and even China are all currently working toward the goal of establishing a long-term presence on the Moon. But the reality is that no matter who gets credit for building the first reliable lunar settlement, it will likely one-day morph into a massive international collaboration unlike anything humans have ever seen. With the ESA and MVA spending time thoughtfully considering the best approaches for building an expandable, adaptable, and most of all permanent lunar base, Moon Village is definitely worth keeping an eye on."[31]

European "togetherness" towards a Moon Village suggests a massive support effort is required. Continuous power will be one of the toughest problems while supplying food, air and temperature control. It will tax any logistics system. Even with use of in-situ materials for 3D printing of structures, massive delivery from the Earth will be needed during the initial phases with a gradual movement to more and more in-situ resource utilization.

---

[31]Moon Village: Humanity's first step toward a lunar colony?
https://astronomy.com/news/2019/05/moon-village-humanitys-first-step-toward-a-lunar-colony

Table 3.3, Reference Destinations - Mars, GEO & Moon

Reference Mission	Metric tonnes to Destination	Comment
Space Solar Power	5,000,000	Power to 12% of Earth's population in an environmentally friendly manner.
Mars Colony	1,000,000	Supporting a Colony with logistics has been underappreciated in the movement off-planet
Moon Village	500,000 estimated	Developing and supporting a colony of residents will require massive movement from Earth

3.3 Interplanetary Mission Support:

Space Elevator support to these three destinations and GEO and beyond will be an important design set of requirements. One of the big questions is how can Space Elevators enable robust movement off-planet to GEO, Moon and Mars? It has become obvious that humans are moving off-planet towards the Moon in this decade and on to Mars after that. This ISEC study must demonstrate that between complementary space access architectures, these missions can be successfully achieved. Rocket portals and Space Elevator infrastructures might be distinctly different; but, they are complementary in their approaches to huge demands supporting Interplanetary Support Missions.

> Remember: This is the transportation story of the 21st century. Reliable, routine, safe, environmentally friendly, and efficient access to space is close at hand.   The combination of transportation infrastructures (rockets and space elevators) will ensure that the missions are supported within global and interplanetary mission needs.

3.4 Large Science Spacecraft to Anywhere:

Science Spacecraft - any size, any where - a future with Galactic Harbours. Future researchers will look at the many tremendous advantages allowed by 1) inherent release velocity at the Apex Anchor and 2) all the mass that is placed in the tether climber is then released towards the solar system destination.   The flexibility of design choices is remarkable with weight not being a major player in design. To place this in perspective, a normal science mission is restricted in design by the tremendous demands of 'lightweight" spacecraft. This is a direct result of the rocket equation and the fact that to raise a science spacecraft to the Earth's Sphere of Influence (SOI) (1 million km radius) 95% of the pad mass is consumed, when using chemical engines.

In addition, the velocity at the SOI is low because to reach a high velocity, it takes rocket fuel burning late in the trajectory (3rd stage), which requires fuel used previously to raise that last fuel to altitude (2nd stage), with more fuel required to raise both sets of fuels off the surface (1st stage). This is called "tyranny of the rocket equation." The rocket equation is insidious. This rocket approach usually results in minimum energy orbits with long flight times for small science payloads.

When launching from the Apex Anchor at 100,000 km altitude, the velocity at the SOI is approximately 7.76 kms/sec. This inherently allows you to go to Mars with no massive rocket required (now, you still need trajectory adjustments and rockets to slow down when reaching destination - same as rocket approach). If you leave from a higher altitude, say 163,000 km Apex Anchor, then you have the velocity at the Earth's SOI to escape the solar system - or to reach any other planet or celestial body in our solar system. Just like the historic approach, you could use gravity assist to increase speeds and approach other planets in shorter times. However, the inherent velocity transferred to the science spacecraft is sufficient to make it to another planet or escape the solar system.

When you now allow any size spacecraft to either rise to or be assembled at the Apex Anchor, you have the remarkable discovery that the science world is no longer restricted by the "light-weight" demands of rockets. The bottom line is:

Any size scientific satellite can go to any planet in our solar system, or beyond when departing from the Space Elevator's Apex Anchor.

3.5 Conclusions:

The unique characteristics of release from Apex Anchor lead to the conclusion that interplanetary flight and movement off-planet requires a complementary infrastructure with rockets and space elevators sharing missions with each system leveraging their strengths. Rocket infrastructures have tremendous strengths; however, the new concept of Galactic Harbours have unique characteristics that ensure they will "enable" interplanetary missions as complementary infrastructure to rockets. These strengths include (see Chapter 4, Arizona State University research):

- Routine Massive Lifts: Each Space Elevator Climber will initially carry 14 metric tonnes of payload to GEO and beyond with departures every day, or 84 Metric tonnes per day (14 x 2 SE x 3 GH) around the globe. This would happen 365 days a year, or 30,660 Metric tonnes per year to GEO and beyond. As the maturity is reached in Galactic Harbours with massive liftoffs, the number moves up to just less that 200,000 Metric tonnes per year to GEO and beyond.

- Routine Daily Lifts: As Space Elevators are designed to lift cargo daily, the release at the other end is parallel. Daily releases towards interplanetary missions will be standard.

- Fast Transits to Mars Available: With the daily release of payloads towards Mars (and other interplanetary destinations) the release from the Apex Anchor imparts tremendous velocity with very little drag from Earth's gravity. As a result, a periodic fast transit to Mars lowers the minimum time to the planet to 61 days.

The dominant question becomes, when looking at it from a Galactic Harbour perspective, "Can we do daily lift-offs with a variety of flight times and distances to Mars?" We not only could go direct, as expected; but, we could go by way of Venus or loop around the Moon for gravity assist to "bend" the orbits. There are so many possibilities — with all the extra velocity, the trips are shorter and can use multiple trajectories. Yes, we can do daily departures towards Mars!

The Galactic Harbour community needs to insert itself into off-planet conversations and express its unique strengths and opportunities. Justification in the past has been focused on Earth missions; however, the discussions should now become movement towards the Moon and Mars. Interplanetary Mission Support is an important future mission for Space Elevators. Galactic Harbours will enable robust mission support off-planet.

These unique needs must be explained within the concept in engineering, financial and programmatic terms. Some conclusions from this analysis could be stimulating!

- Galactic Harbours will enable missions to Moon and Mars!
- Only Space Elevator transportation infrastructures can deliver the requirements of logistics equipment and supplies to the Moon and Mars in a timely manner.
- Colonization on Mars cannot happen without the logistics support of Space Elevators!
- Launch Windows to Mars every 26 months - be Damned! [my favorite]

Reliable, routine, safe, environmentally friendly, and efficient access to space is close at hand. The Space Elevator is the Galactic Harbour, and an essential part of the global and interplanetary transportation architecture. The requirements (needs) that are being discussed today can be fulfilled with a Galactic Harbour infrastructure enabling the movement beyond Earth throughout this new century. One realization that has not been recognized is future planning does not require building everything on-site. In-situ resources are necessary, but with Space Elevators the delivery of cargo, supplies, habitats, water, air and all future needs can continue to be procured from the Earth through routine, inexpensive, reliable and daily deliveries. The three reference mission destinations (Mars - 1,000,000 Metric tonnes, Moon Village - 500,000 Metric tonnes, and Space Solar Power - 5,000,000 Metric tonnes) will be enabled by an Interplanetary Transportation Infrastructure called the Galactic Harbour. This is accomplished with three elements necessary to have robust interplanetary mission support: (1) massive movement of cargo, (2) daily liftoffs, and (3) periodic rapid transit.

This is the transportation story of the 21st century.

Chapter 4 - Arizona State University (ASU) Research[32]

4.1 Background:

The envisioned Missions to Mars and the Moon are of incredible size and scope as well as exciting and stimulating. If you have ever worked the problem of movement to Mars you would have talked about:

- Earth's sphere of influence [1 million kms]- Mars' sphere of influence [600,000 kms]
- 'Launch Windows' only every 26 months due to the need for Hohmann transfers [minimum energy path] and the position of Earth/Mars around the Sun.
- Heavy rockets needed to get the energy to: reach LEO [9 ++ km/sec] plus Mars transfer trajectory [3.6 km/sec or so]
- Velocity at the Earth's sphere of influence enabling flights to Mars [2.94 km/sec at 1 million kms]
- Added to Earth's velocity, sufficient to go into Hohmann transfer towards Mars
- Roughly 5 km/sec to slow down and go into circular orbit around Mars [400 km altitude] [space elevator "tossed" orbital approaches are similar in propulsion needs along the orbit and at the destinations]

Several of the "big questions" are:
- Question: "How often can we expect to deliver payloads to Mars from Space Elevator's Apex Anchors?" What are the launch windows? - weekly or daily?
- Question: "What are the transit times when departing the Earth's sphere of influence with three times the energy of past chemical rockets? [non-traditional approaches - Lambert's Solution]
- Question: "With those two discussions initiated, the concept of multiple metric tons of mission support logistics arriving at Mars (or the Moon) per day is remarkable. How will that change the approach to support human movement off-planet?

---

[32] Note: Much of this chapter was written for a paper by J. Torla and Associate Professor Peet. The ASU team members were: Mars Team Members: Mark Lyons, Runa Nakamur, Renzo Curay De La Ros, Shawn Michae Bauer, and Nathan Renard, and Io Team Members: Nicholas Iannacone, Tyler Mebane, Avi Brahmbhatt, Samuel Bolar, Jose Valenzuela, Jonathan Johnson, and the team leader Ryley Miller.

The ASU study was to address: How does one assess the strengths of Galactic Harbours' logistics support for Interplanetary missions? A few of the questions inherent in this approach are:

- What speed are you going at the sphere of influence of Earth when released from Apex Anchor at an altitude of 100,000 km?   = 7.76 km/sec
- When you are going that high velocity (approx. 3 x traditional Hohmann approach) at the sphere of influence, how long does it take to get to Mars?
- Oops, you must rearrange your thinking as Hohmann transfers are NOT in play. The payloads are in an elliptical orbit around the sun at much great speeds.
- Assumptions: Six operational space elevators; each able to "raise" 14 Metric tonnes of payload per day (individual missions could vary in mass); and, late 2030's to early 2040's operational. In addition, growth in capacity will be discussed as mission needs increase and technologies are enabled.

12.35 km/sec
163,000 km

11.4 km/sec
150,000 km

7.76 km/sec
100,000 km

3.078 km/sec

Geosynchronous
Altitude

Figure 4.1, Space Elevator
Launch Geometries[33]

The question becomes, when looking at it from a Space Elevator perspective, "Can the Galactic Harbour infrastructure do daily launches with a variety of flight times and distances to Mars?" The study results show that the spacecraft not only could go direct, as expected; but, could also go by way of Venus or loop around the Moon for gravity assist to "bend" the orbits. There are so many possibilities — but with all the extra velocity, the trips are shorter and can leverage various orbits. What does it mean with respect to the big picture of supplying logistics support to off-planet robotic and human exploration and colonization? One interesting area of study, for the future, must include the planetary incoming velocity and how to slow down for rendezvous and landing? As there are so many questions to be answered, ISEC leveraged a relationship with Arizona State University and Associate Professor Matthew Peet leading a team of students. Their research into the puzzle is discussed during the rest of this chapter.

---

[33] Torla, James and Matthew Peet, "OPTIMIZATION OF LOW FUEL AND TIME-CRITICAL INTERPLANETARY TRANSFERS USING SPACE ELEVATOR APEX ANCHOR RELEASE: MARS, JUPITER AND SATURN," International Astronautics Congress (IAC-18-D4.3.4), Washington D.C., 2019.

4.2 Introduction:

Associate Professor Matthew Peet and Dr. Peter Swan of ISEC initiated research in the spring of 2019 assessing the strengths of releasing payloads at the Apex Anchor for interplanetary missions. The students, lead by James Torla, examined the potential impact of a Space Elevator Apex Anchor release of cargo for permanent human habitation on Mars and the Moons of Jupiter and Saturn.34 The students constructed a set of trajectory simulation codes, in a coding language (Matlab), determining flight trajectories made available by the construction of a Space Elevator and release at its Apex Anchor. The code explores the practicability of interplanetary launches beyond restrictions imposed by the optimal Hohmann transfer methodology. Furthermore, the code was able to determine w[35]hen, if ever, a Space Elevator had the capacity to deliver a craft to Mars at fuel budgets and times of flight more efficiently than the ideal Hohmann transfer window. Apex Anchor release trajectories refer to the low-cost interplanetary insertions corresponding to the initial velocity vector achieved at the Apex Anchor. The velocities in most cases were beyond Earth escape velocities and under certain conditions could yield interplanetary transfers with minimal Delta-v requirements. In this work, the students used iterative methods based on a variation of Lambert's Problem to determine the minimal Delta-v direct transfer from an Apex Anchor to Mars under a variety of initial conditions and time-of-flight constraints. The Lambert Problem, entering the sun based elliptical orbit on the side, is shown in Figure 4.2.

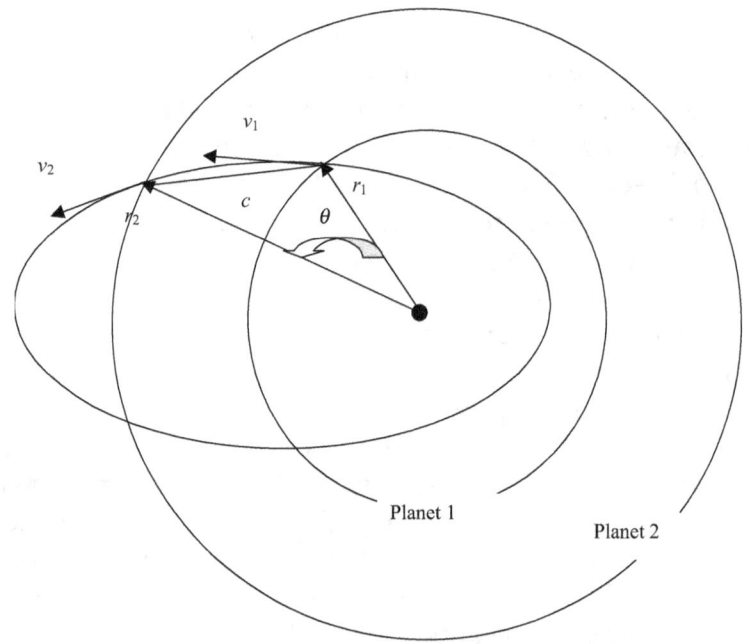

Figure 4. 2, Lambert Problem Geometry

Permanent human habitation of distant planets requires both cost-effective methods of transportation of a

34 Torla, James and Matthew Peet, "OPTIMIZATION OF LOW FUEL AND TIME-CRITICAL INTERPLANETARY TRANSFERS USING SPACE ELEVATOR APEX ANCHOR RELEASE: MARS, JUPITER AND SATURN," International Astronautics Congress (IAC-18-D4.3.4), Washington D.C., 2019.
35 Kemble, Stephe,, Interplanetary Mission Analysis and Design, pg 9, Springer, 2006

massive amount of materiel and the ability to reduce the time-of-flight for human passengers and safety-critical supplies. The student results demonstrated that the use of an Apex Anchor release can address both needs by dramatically reducing the time-of-flight for a fixed Delta-v budget; or, conversely, dramatically reducing the Delta-v budget when time-of-flight constraints are relaxed.

In addition, Galactic Harbours are a strong contender to provide the low cost launch options necessary for the continued growth of space flight. Due to the rotation of the Earth, any object released from the Apex Anchor would have the necessary velocity for interplanetary transit on a hyperbolic orbit escaping Earth. This excess velocity also negates the use of a Hohmann transfer and could permit the release of spacecraft at times throughout the synodic period. The concept is shown in Figure 4.1, Apex Anchor Velocity. As shown in this figure, Space Elevators have a base fixed at the equator and as such the rotation rate of the earth at 360 degrees per day. Thus the Apex Anchor of length 100,000km will have a velocity of 7.76 km/s (and 11.4 km/sec when the Apex Anchor is at 150,000 km altitude - an option not pursued inside the ASU study. In addition, future research will look at the 163,000 km altitude where Solar System escape velocity is reached.). Therefore, the velocity of any object released from the Apex Anchor in the Earth Centered Inertial (ECI) frame would already be well beyond Earth's escape velocity. Furthermore, there are times when the excess velocity allows for transit to Mars using free release – with no additional Delta V requirements.

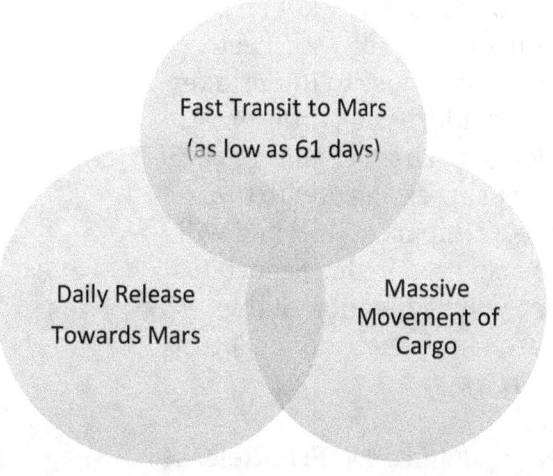

Figure 4.3, Newly Recognized Strengths

The Galactic Harbour does have some disadvantages. The Space Elevator is attached at the equator rotating within the equatorial plane and not the Solar System's ecliptic plane. Thus either Mars at arrival must also lie in the equatorial plane or there must be a plane change of up to 23 degrees during the flight. This traditional flight correction complicates the analysis of benefits. To date, however, this question is not well understood and is part of the study analysis. There have been no quantitative studies of the Delta v savings and launch windows enabled by the development of Space Elevators to compare with. The purpose of this chapter, then, is to assume a Space Elevator of length 100,000 km and quantify the Delta v requirements for Mars transfer and orbital insertion as a function of desired launch date. This ASU research broke the discussions down into two approaches looking at various aspects of the problem and then tied them together in the summary. First, the students explored the question of transfer to Mars using free release; or. only Delta v for an orbital plane change depending on the position of Mars at arrival. This analysis did include the Delta v needed for orbital capture at Mars on arrival and simply calculates the release time when a free release trajectory will intersect with Mars. The second approach explored the minimum Delta v trajectory including Delta v for both launch and orbital insertion at Mars as a function of date. This approach searches over a set of flight times, calculating the minimum Delta v required for Lambert transfer over all possible release times for a given day, then saved it. These results demonstrated that space elevators significantly reduce the total Delta v required for Mars transfer and insertion over any conceivable Earth-based rocket launch scenario.

4.3 Method:

The students were challenged in their class to develop the code for understanding this new capability of release from the Galactic Harbour's Apex Anchors. Most of the work shown in this study report deals with the mission to Mars; however, the students also looked at going to Jupiter's moon, to Io. In addition, in the spring of 2020 a new set of students from Professor Peet's orbital class chose to undertake research addressing different challenges. These challenges are slowly progressing due to the Covid-19 impact at ASU and will not be covered in this report (there will be mention of this research in the section on Future Research).

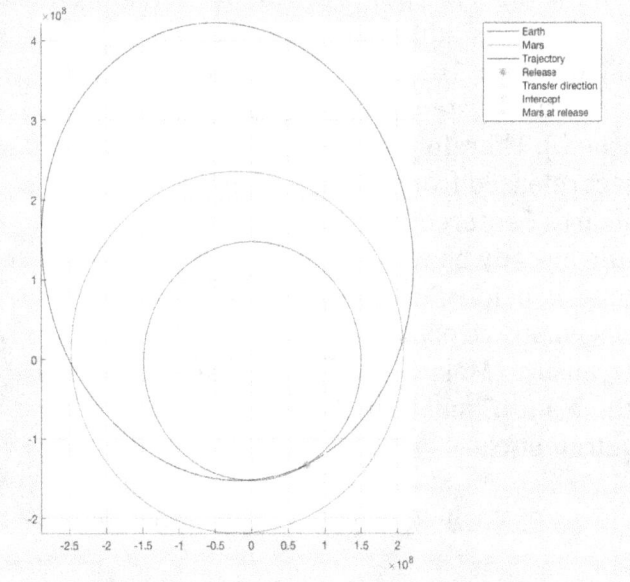

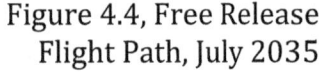

Figure 4.4, Free Release
Flight Path, July 2035

By constructing a set of trajectory simulation codes, in Matlab, the ASU researchers determined flight trajectories available through the construction of Space Elevators. Their code explores the practicability of interplanetary launches at times well beyond the limitations of the optimal Hohmann transfer window. The students also determined when, if ever, a Space Elevator has the capacity to deliver a craft to Mars at fuel budgets and times of flight significantly more efficiently than the ideal Hohmann transfer window. Figure 4.4, Free Release Flight Path, July 2035, shows the non-Hohmann Transfer approach from Earth (smaller circle) to Mars (larger circle) in an optimum orbit along an ellipse of immense energy. The ellipse can "cut the corner" because it has high velocity. There were three key things that needed to be determined to bound the problem leading to full understanding.

- The length of the space elevator is critical as it provides the velocities necessary for interplanetary travel. The students selected 100,000 km as described by the International Space Elevator Consortium in the Space Elevator Architecture and Roadmaps (see Appendix D).
- This length produces a velocity of approximately 7.76 km/s in the Earth's ecliptic plane.
- This velocity, combined with the Earth's motion around the sun, provided a first boundary condition for simulation; the heliocentric launch velocity of the spacecraft as a function of time (vector summation).

The second boundary condition occurs at the spacecraft launch? From the launch date, comes the relative location of the planets during their synodic period leading to the determinant for the time of flight and the delta V required for interplanetary travel.. There is a synodic period for Earth and Mars of approximately 26 months during which the two planets get as close as 54.6 million km and as far away as 401 million km. Although, neither planet has a perfectly circular orbit, the eccentricities are small and were discounted for the purposes of ease of calculation. The researchers picked the synodic period beginning with 1 January 2035 and ending with 28 February 2037; around when the first Space Elevator operations might begin.

The third and final determination necessary to begin the code construction was how often should the launches become. A significant strength from Galactic Harbours is the elevators has reliable infrastructure that can be used each day at any time. While it is true that the Space Elevator can deliver objects to Earth's orbit at all times, the precise timing for the release of the spacecraft for interplanetary missions matters a great deal since it only takes four minutes for the Earth and the Space Elevator to rotate by 1 degree. To keep things simple, while maintaining a small enough sample that the computational requirements for the code would not exceed a reasonable limit, four minutes for each launch opportunity was chosen.

With all of the initial boundary conditions decided, the code was constructed. The code was made up of three separate scripts each with a slightly different job to perform.
- The first script served to track the spacecraft from its initial launch trajectory to determine when –If ever– it intercepts Mars and what Delta Vs and time of flights it may take to get there. The Delta Vs accounted for in this code are only those strictly necessary to account for the plane change from the Equatorial to the Ecliptic and for capture by Mars.
- The second code utilizes Lambert's Problem to find the absolute minimum Delta V permitted every single day for flight durations up to 365 days.
- The third and final script was simply a data processing script that took the received results from the other two scripts to plot them separately and jointly.

4.4 Free Release Script:

The development of the script then approaches the problem sequentially.

- The first script begins by setting a couple constants and operating parameters like the time period or the simulation for the length of the elevator. One key operating parameter is that of Phase Change 'PC' or No Phase Change 'NPC'. This variable is used to moderate whether a burn is made to account for the plane change from the Equatorial plane to the Ecliptic. When NPC is selected the code will only return the flight paths where the spacecraft intercepts Mars solely due to the initial release velocity.
- The first computation done is to find the planetary elements of both Earth and Mars at the given launch date and time using J2000 data base[36]. From there, the velocity of the spacecraft can be calculated from the Earth's initial orbital elements and the length of the elevator. The position and velocity vectors in the heliocentric frame are then converted to heliocentric orbital elements.
- The Delta V necessary to account for the plane change from Equatorial to the Ecliptic was then calculated using the law of sines with an angle change determined by angle between the heliocentric velocity vector and the orbital plane. The magnitude of the velocity vector is not changed during this operation. Once the transfer orbit has components solely in the Ecliptic plane, the orbit is propagated using the polar equation out to the Martian orbital radius. The code then checks whether the craft is within Mars' sphere of influence at that point in time. When it is, the flight path is a successful intercept, the Delta V due to intercept velocity at Mars is calculated using the Oberth effect. This process is repeated for each possible release time on each day within the given search parameters.
- The summary shows time of release, time of arrival (thus time of flight), and delta velocity necessary for several successful orbital paths per day.

4.5 Lambert's Method Script:

The Lambert's initial setup is similar to the Free-release script.

- For each possible release time, the code calculated the initial heliocentric velocities of the departure and target planets. Then the velocities were used to determine the minimum elliptic Time of Flight (ToF) for the Lambert's problem. Recall that Lambert's problem is: given two position vectors and a TOF, to determine the corresponding initial and final velocity vectors. Our algorithm implements the solution of Lambert's problem using the method and a Matlab code (with minor modifications) developed by Richard Battin. This yields, for a given position of Earth, and TOF, the velocities at both Earth and Mars that are required to intercept Mars for the given TOF.

---

[36] The currently-used standard epoch "J2000" defined by international agreement - from wiki, 28 June 2020.

- The search for the optimal Time of Flight is discretized using 1/10 of a day from the minimum elliptic TOF to an arbitrary maximum of 365 days (orbital paths reached beyond this maximum, and were not shown). The proposed position vector of Mars at intercept for each TOF is calculated using standard orbit propagation. For each release time and TOF, the solution to Lambert's problem then yielded an initial and final velocity vector.
- The first Delta V for the flight is now calculated by finding the difference between the initial velocity and the velocity of the apex anchor. The second Delta v is found by finding the difference between the velocity at arrival and the velocity of Mars. This relative velocity is then propagated in the Martian sphere of influence to a periapse equal to the Martian radius and the Delta v required for transition to highly elliptic (e=1) orbit is calculated. The initial and final Delta v's are summed and the minimum Delta v for each proposed launch date is calculated. The minimum Delta V found for the launch times and TOF for each launch date are then saved in a cell array along with the angle at release, the time of launch and interception, and the Time of Flight.

4.6 Data Processing Script:

The use of a "heat map" illustrated the key elements of the analysis to include release date, time of flight (departure and arrival dates), and velocity required. The data processing script simply loaded that saved data files from the previous two scripts, converting the cell arrays to Matlab tables, and then producing heat maps of Delta V for TOF and Launch date from the tables. These heat maps only account for the existent intercept data, so the time of launch axis may not be linear. However, each chart does show the relative magnitude of Delta V required for each Mars interception, though the scales are not constant between charts. There are four heat map charts produced in total;
- one charts the Free Release trajectories calculated in the first script (Figure 4.4):
- another charts the same Free Release trajectories in greater detail for the month of July, 2035 (Figure 4.5)
- another charts the Lambert interceptions from the second script (Figure 4.6), and
- the last combines the intercept data from both the Lambert's and the Free Release scripts into one chart (Figure 4.7).

4.7 Results:

A standard mission to Mars during the Hohmann transfer window in 2035, as described in NASA's Interplanetary Mission Design Handbook, requires a Delta V of approximately 12 km/s to arrive at Mars within 200 days. During the same time period, with a No Phase Change free-release trajectory (Figure. 4.4), a Space Elevator can allow a spacecraft to rendezvous with Mars in 170 days with a minimum delta V of 3.7 km/s. The Space Elevator can also deliver a craft to Mars within 76 days with 10 km/s of Delta V during the same time period that's highlighted in Figure. 4.5.

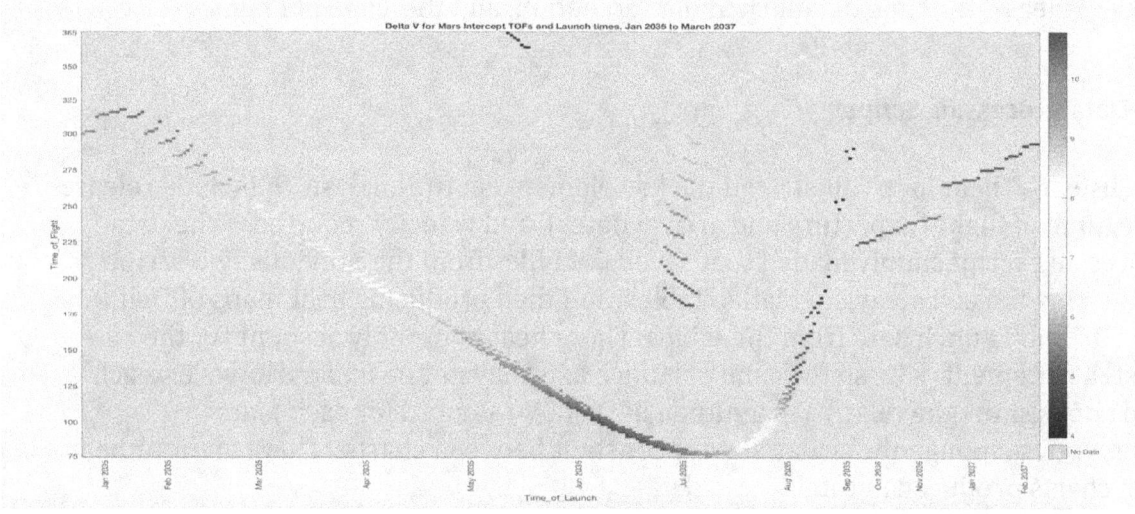

Figure 4.5, Delta V Heat Map for Free Releases, Jan 2035 to Mar 2037
[left axis - ToF 60 - 365 days; right axis - delta v req. 4 to 11 km/sec; horizontal axis - date of launch Jan 2035 to Feb 2037; points on graph designate ToF while hight designates "cost in rocket fuel" - Low point shows 78 days, medium energy]

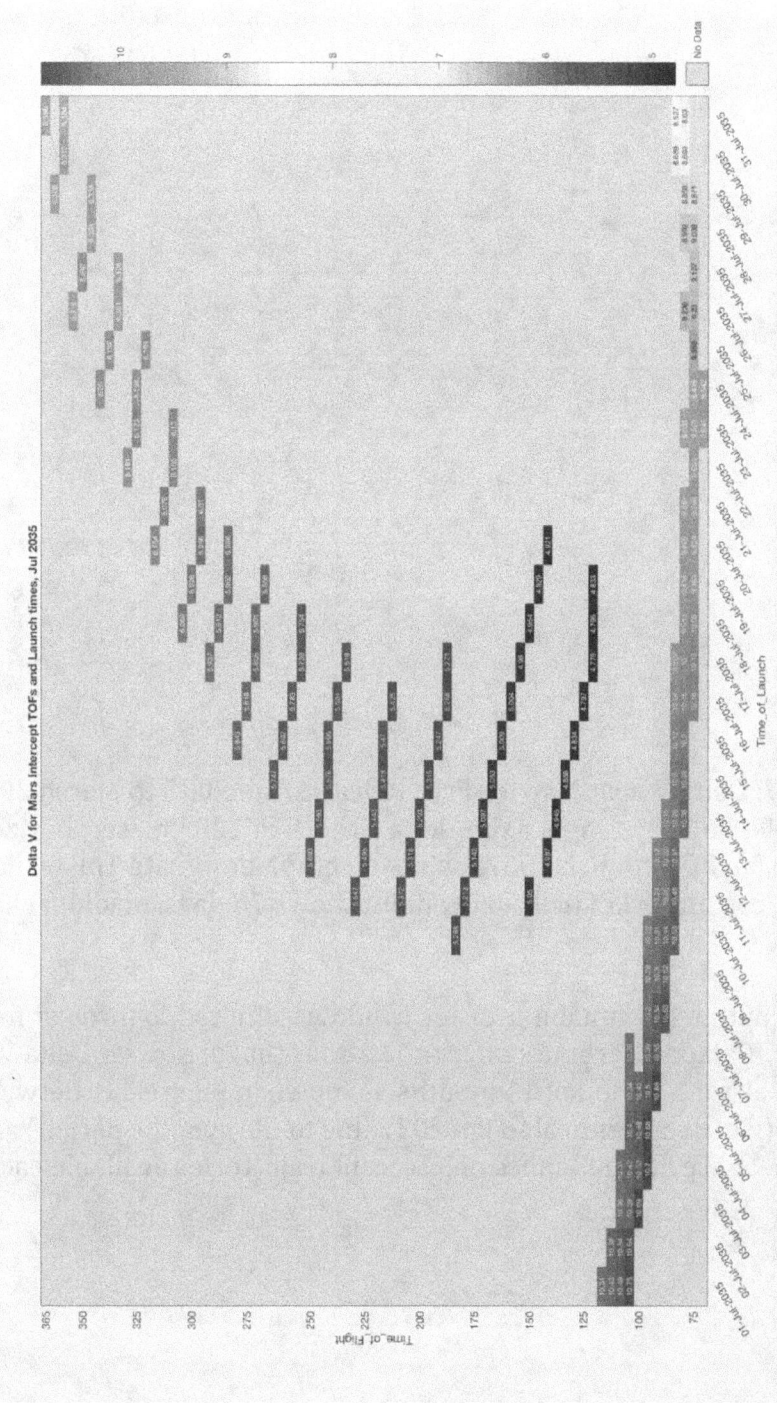

Figure 4.6, Delta V Heat Map for Free Releases for July 2035
[left axis - ToF 75 - 365 days; right axis - delta v req. 4 to 11 km/sec; horizontal axis - date of launch July 2035; points on graph designate ToF while height designates "cost in rocket fuel" - Low point shows 78 days, medium energy - also shows that for certain days, there are multiple paths to get to Mars with varying times and fuel]

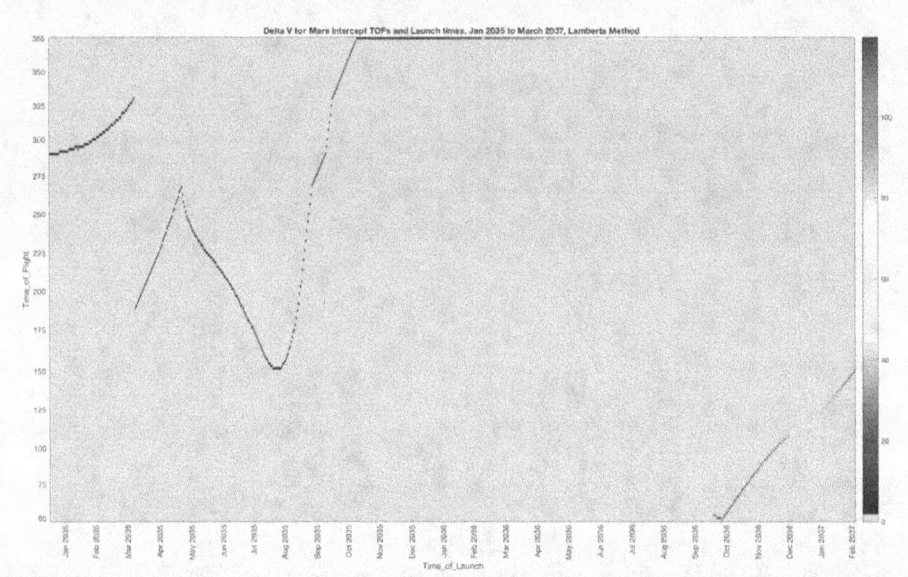

Figure 4.7; Delta V heat Map for Free Releases. Jan 2035 to March 2037
[left axis - ToF 60 - 365 days; right axis - delta v req. 0 to 120 km/sec; horizonal axis - date of launch Jan 2035 to Feb 2037; points on graph designate ToF while height designates "cost in rocket fuel" - Low point shows 78 days, medium energy]

In fact, while the optimal Hohmann transfer window is limited to three or four weeks every 26 months, Galactic Harbours offer trajectories that are lower Delta V and quicker every week for a period of 11 months as shown in Figure 4.8. Between November 2034 (which is identical to Jan 2037 due to the synodic period) and October 2035, there are an abundance of intercept trajectories available each week.

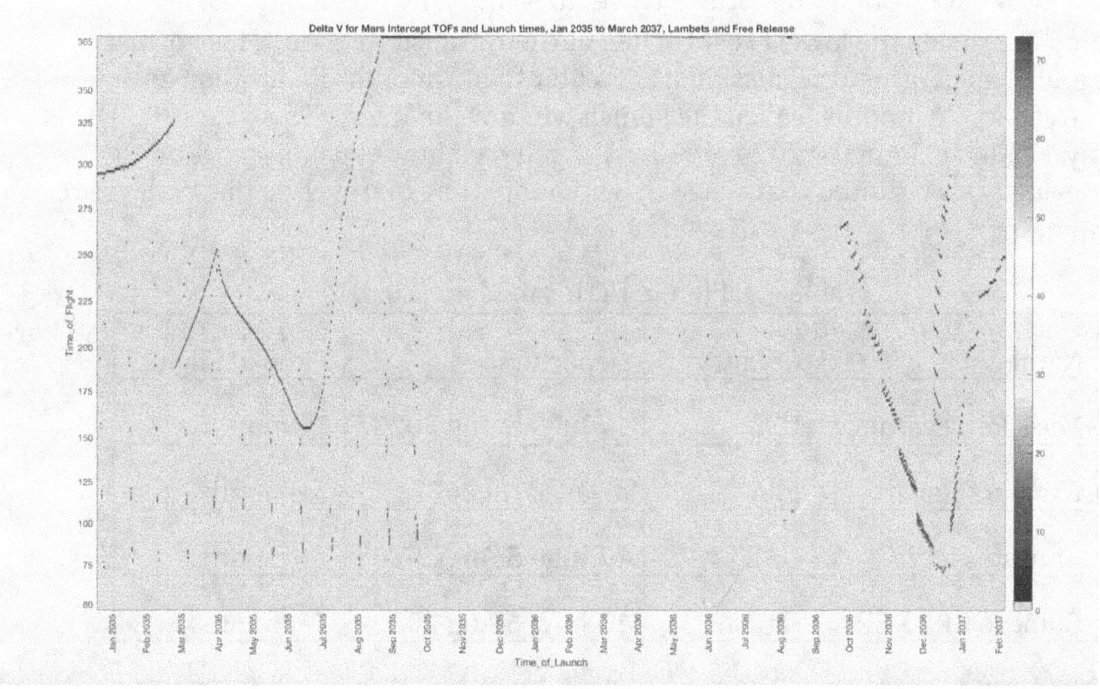

Figure 4.8; Delta V heat Map for Lambert's Script. Jan 2035 to March 2037
[left axis - ToF 60 - 365 days; right axis - delta v req. 0 to 75 km/sec; horizonal axis -
date of launch Jan 2035 to Feb 2037; points on graph designate ToF while height
designates "cost in rocket fuel" - Low point shows 78 days, medium energy]

4.8 Conclusions:

To place the conclusions in perspective from this research conducted at the Arizona
State University, a quick look at historic approaches to Mars is necessary. The first
thing to realize is that there is a twenty-six month delay between launch
opportunities for rocket propulsion from Earth to Mars. This is a natural trade space
across orbital positions, energy required and time of flight resulting. Some of these
numbers are shown next:

4.8.1   Quick Background on Historic Travel to Mars. The Hohmann transfer is designed to provide the lowest cost (in fuel burned) transit to Mars. This normal Hohmann Transfer (historic missions) is greater than 7 months in duration and launches every 26 months (called the launch window for Mars). This year (2020) it is roughly 17 July to 5 August 2020 (19 days). As the window expands, more energy (delta velocity) is required to reach Mars with more time of travel, so the 19 days are optimum.

Table 4.1, Planned Trips to Mars (2020)

Name	Country	Departure	Time of Flight
Mars Perseverance	USA	17 July- 5 Aug 2020	7+ months
ExoMars Rover	ESA	delayed till 2022	7+ months
Emirates Mars	UAE	17 July- 5 Aug 2020	7+ months
Chinese GRSO-SR	China	17 July- 5 Aug 2020	7+ months

This year (2020) will feature three attempts to rendezvous with Mars from three nations. (note, the ESA/Russian mission was delayed until next window -26 months later) These launches will be scheduled during the open window and will travel roughly seven months to Mars for the mission.   Each will leverage large rocket propulsion and have extreme restrictions on weight as the trip is expensive in fuel. Examples of past missions are shown in Table 4.2.

Table 4.2, Past Trips to Mars

Name	Departure	Time of Flight
Maven	18 Nov 2013	9+ months
ExoMars	14 Mar 2016	7+ months
Insight	5 May 2018	7 months

4.8.2 Comparison to Space Elevators: The research conducted by ASU on the topic of time of flight and launch opportunities showed some surprising results. These will be broken down into two categories: Time of Flight and Launch Window. During the following discussions, care must be taken to consider the calculated delta velocity to have a successful flight and orbital capture at Mars. The next two charts in this section show various characteristics of flight shown to include required delta velocity.

The full calculations were done for the two years between rocket launch windows to Mars (2035-2037), but the full data set will not be shown as it is too large. Selected sections will be chosen to show the optimum trips to Mars. A sample is shown next.

Table 4.3, Trips to Mars from Free Release Data Technique

Time of Launch	Time of Intercept	Time of flight (days)	Delta V (Km/sec)
13-Oct-2036	13-Dec-2036	62	31.69109344
14-Oct-2036	14-Dec-2036	62	29.52264214
15-Oct-2036	15-Dec-2036	62	27.68490601
16-Oct-2036	16-Dec-2036	61	26.16316414
19-Oct-2036	18-Dec-2036	61	22.94289017
20-Oct-2036	19-Dec-2036	61	22.72153091
21-Oct-2036	20-Dec-2036	60	22.69838524
10-Apr-2035	8-Oct-2035	181	4.015966892
11-Apr-2035	10-Oct-2035	182	4.001704216
15-Apr-2035	18-Oct-2035	186	3.956878901
16-Apr-2035	20-Oct-2035	187	3.941791534
17-Apr-2035	22-Oct-2035	188	3.938188076
18-Apr-2035	24-Oct-2035	189	3.928704977

Table 4.4, Trips to Mars from Lambert Approach

Time of Launch	Time of Intercept	Time of flight (days)	Delta V (Km/sec)
15-8-2035	20-Nov-2035	97	5.752464338
15-8-2035	22-Nov-2035	99	5.693665943
15-8-2035	23-Nov-2035	100	5.571570388
16-8-2035	4-Jul-2036	324	6.502765567
16-8-2035	24-Nov-2035	100	5.552539488
16-8-2035	25-Nov-2035	101	5.491620887
16-8-2035	1-Dec-2035	108	5.194464684
29-7-2035	15-Oct-2035	78	8.90657348
29-7-2035	14-Oct-2035	77	8.854903468
29-7-2035	14-Oct-2035	77	8.88298226
29-7-2035	14-Oct-2035	78	8.866683752
29-7-2035	14-Oct-2035	78	8.789739678

By leveraging two approaches to the calculations for the trips to Mars, the ASU research has shown that there are tremendous advantages from leaving at the Apex Anchor with an incredible initial velocity. The charts above show the variations in time of flight, date of launch and energy required. They encompass launches across the normal "non-launch" period for rockets and have many opportunities with less energy and definitely less travel time. Perhaps the main benefit of the Space Elevator for trips to interplanetary destinations is in the broad range of possible departures with varying flight times. Furthermore, the Space Elevator launch times contain a great deal of flexibility within each time period. Figure 4.5 shows the Free Release Delta V heat map for July of 2035, which is the month with the most intercepts with Mars. During this month there are a great number of different launch types available every day from the Space Elevator. On July 16th, for example, there are 3 separate low time of flight launch times and 6 separate low Delta V launch times. Different clients with different time windows and budgets could purchase the launch time that best fits their needs with the greater flexibility offered by Galactic Harbours.

Time of Flight (ToF): The normal rocket approach is launch only every 26 months, adapting to the orbital positions of the planets. As such, the earlier charts showed the past few missions and the upcoming missions in the summer of 2020. As shown in the heat charts and the tables above, the time of flight from Space Elevators varies from 61 days to over 365 days. A key result of this research is that the planner may release cargo to go to Mars any day of the year across the rocket restricted 26 month launch window . As shown, the faster transits have some higher thrust requirements, but many are low delta velocity as well. Two optimum examples are compared to the shortest time of flight with its high-energy requirement:

Table 4.5, Optimum Examples of Energy vs. ToF

Time of Launch	Time of Intercept	Time of flight (days)	Delta V (Km/sec)
15-8-2035	20-Nov-2035	97	5.752464338
29-7-2035	14-Oct-2035	77	8.88298226
18-Oct-2036	17-Dec-2036	61	23.38561821

Release Dates: The beauty of this approach of releasing from the Apex Anchor at an initial velocity is that there is no built in restriction of when to release towards Mars similar to the 26 month launch window restrictions for rockets. The calculations have shown, in both the Lambert approach and the free release approach, that the Apex Anchor release could occur any day of the year. The difference is that there are paths that would take longer than 365 days to reach Mars or would require very large delta velocities to both change planes and slow down at Mars. This flexibility in release dates enables the logistician to plan for delivery of cargo to Mars as needed vs. every 26 months. Some trips are rapid, some trips require large delta velocities, and some trips are optimum for both. The supply of the colony on Mars can be planned and executed over the two years between rocket launch opportunities carrying people.

Energy Required: For rockets, this insertion into Mars Transfer orbit requires 3.6 km/sec after arriving in Low Earth Orbit with approximately 9.3 km/sec, or roughly 13 km/sec of velocity through fuel consumption. This does NOT include trans-Mars ecliptic plane corrections nor the fuel to slow down the vehicle to be captured by the planet. The numbers shown in the above charts include both and reflect the full energy to leave the Apex Anchor, adapting for ecliptic plane differences and then slowing down for inclusion in a elliptical orbit around Mars. As shown above, the energy required is dependent on the location of the two planets when departure occurs and the time of flight requested. There are several situations where the numbers for delta velocity are optimized and the time of flight is low.

> NOTE: The delta velocity is only reflective of the change in velocity required and is a "mass-less" unit. When one compares a number from the ASU charts, the reader must recognize that when mass is included, the ASU numbers only reflect payload mass acceleration. When one calculates the mass for a rocket mission to Mars, first one must multiply the 9.3 delta v requirement times the mass of the full rocket to show the fuel required to get off the planet and then 3.6 km/sec times the mass of the payload going to Mars. The amount of fuel for this case is enormous. After the Earth rocket example reaches the SOI, then the comparison is one to one - mass of payload going to Mars is accelerated by fuel consumed. The difference is that payloads from the Apex Anchor go up on an infrastructure by electricity, and once released from the Space Elevator require fuel consumption to correct for inclination and then slow down for capture at the target planet.

Overall conclusions from the ASU research:

- Interplanetary travel is greatly enhanced by release from the Apex Anchor in time of flight, required delta velocity and launch opportunities.
- Release velocities at the Apex Anchor expand the freedom of launch for both scheduled and emergency responses.
- Low cost options exist when looking at the lower delta velocity opportunities
- Specialized missions can be designed based upon predictable orbital positions and the rotation of the Apex Anchor around the Earth. Low time of flight missions are especially appreciated by mission directors and missions in emergency situations.
- Some details:
  - ToF - greater than 69% are less than 200 days with delta v's between 4 and 14 km/sec. Many ToFs are just beyond 60 days. Over 300 flights have less than 3 months of ToF.

  o Delta Velocity - 74% have less than Hohmann transfer of 13 km/sec(1,500 options). 16 of 26 months have many low delta velocity options. Over 1,000 releases have less than 9.4 km/sec required. 17 release require less than 1 km/sec with ToF of 170 days.

  o Launch Windows - There are four weeks to 62 weeks of optimum flight paths across the 26 month rocket launch window with reasonable ToFs and delta velocities. Essentially, there are daily releases with various options for ToF and delta velocities.

4.8.3 Additional Destinations: As discussed in previous chapters, the flexibility to support various missions to diverse destinations is a tremendous strength of Galactic Harbours. The support to the various missions is made possible by the tremendous flexibility in release velocities (and energy). The next chart summarizes a variety of altitudes associated with various architectures for Space Elevators (historic). This 2019 ASU study uses 100,000 km altitude, or 106,378 km radius from center of Earth to Apex Anchor. With this altitude (or potential energy) and tip velocity (or kinetic energy), the release towards the Moon or Mars would be 7.76 km/sec. In addition, the various velocities at release from the Apex Anchor can enable flights to all the destinations of choice within our solar system (and even beyond with the proper additional propulsion).

Table 4.6: Height of Apex Anchor Alternatives

Length (kms)	Author	Speed (Km/sec)
96,000	Obayashi (2013)	7.466
100,000	Edwards (2002), IAA (2013 & 2019)	7.757
120,000	Knapman (2019)	9.216
144,000	Pearson (1975)	10.966
150,000	ASU alternative (2018)	11.403
163,000	Solar System Escape Vel (2020)	12.352

4.8.4 Bus Schedules: One of the surprising conclusions from this study is that a new concept was developed - a bus schedule to Mars (and other interplanetary destinations). Figure 4, Mars Bus Schedule, shows an example of departure days (month of July 2035) towards Mars with flight times shown (weekly dates were chosen to show - in reality they are daily). When you have a permanent transportation infrastructure in place, the scheduling of asset movement is routine - like taking a bus to Mars.

Research at Arizona State University, sponsored by Associate Professor Matthew Peet and led by James Torla, produced a conclusion that mission support from Galactic Harbours is an enabling factor in humanity's expansion off-planet.

Table 4.6, Bus Schedule to Mars 7/1/2035

This is the transportation story of the 21st century. Reliable, safe, and efficient access to space is close at hand. The Space Elevator is the Galactic Harbour, and an essential part of the global and interplanetary transportation infrastructure.

Bus Schedule for Interplantary Transportation
when departing from Galactic Harbour Apex Anchor

Bus Schedule, from Apex Anchor 2035

Date	Departure	Destination	Flight Time	Arrival	Comments
7/1/2035	Indian #1	Mars	87 days	9/26/2035	
7/1/2035	Pacific #1	Mars	86 days	9/25/2035	
7/1/2035	Pacific #2	Mars	84 days	9/22/2035	Fast

Bus Schedule, from Apex Anchor 2035

Date	Departure	Destination	Flight Time	Arrival	Comments
7/8/2035	Indian #1	Mars	81 days	4/14/2035	
7/8/2035	Indian #2	Mars	81 days	4/14/2035	
7/8/2035	Indian #1	Mars	80 days	4/13/2035	Fast

Bus Schedule, from Apex Anchor 2035

Date	Departure	Destination	Flight Time	Arrival	Comments
7/15/2035	Indian #1	Mars	79 days	10/2/2035	
7/15/2035	Indian #1	Mars	79 days	10/2/2035	
7/15/2035	Indian #2	Mars	79 days	10/1/2035	
7/15/2035	Indian #2	Mars	79 days	10/1/2035	
7/15/2035	Pacific #1	Mars	78 days	9/30/2035	Fast
7/15/2035	Atlantic #1	Mars	190 days	1/21/2036	
7/15/2035	Atlantic #1	Mars	182 days	1/13/2036	
7/15/2035	Atlantic #2	Mars	173 days	1/4/2036	
7/15/2035	Atlantic #2	Mars	164 days	12/25/2035	
7/15/2035	Atlantic #1	Mars	154 days	12/15/2035	

Bus Schedule, from Apex Anchor 2035

Date	Departure	Destination	Flight Time	Arrival	Comments
7/22/2035	Pacific #2	Mars	77 days	10/7/2035	Fastest
7/22/2035	Pacific #2	Mars	77 days	10/7/2035	Fastest
7/22/2035	Pacific #1	Mars	223 days	3/1/2036	

Bus Schedule, from Apex Anchor 2035 to Moon

Date	Departure	Destination	Flight Time	Arrival	Comments
every day	Indian #1	Moon	14 hours	+ 14 hours	
every day	Indian #2	Moon	14 hours	+ 14 hours	
every day	Pacific #1	Moon	14 hours	+ 14 hours	Fast
every day	Pacific #2	Moon	14 hours	+ 14 hours	
every day	Atlantic #1	Moon	14 hours	+ 14 hours	
every day	Atlantic #2	Moon	14 hours	+ 14 hours	

4.8.5 Future Research: The past semester students (2019 & 2020 classes) were challenged to develop the code for understanding this new capability of release from the Galactic Harbour's Apex Anchors. Most of the work shown in this study report deals with the mission to Mars; however, the students also looked at going to Jupiter's moon, IO. In addition, in the spring of 2020 a new set of students from Professor Peet's orbital class chose to undertake research addressing different challenges. These challenges are slowly progressing due to the Covid-19 impact at ASU and will not be covered in this report. As this ISEC study report is closing, the planning for Fall of 2020 and Spring of 2021 has started. Some ideas that could be pursued are:

- Apex Anchor to Mars (with a receiving Space Elevator rendezvous)
- Apex Anchor to the Moon (with a receiving Space Elevator rendezvous)
- Trips to the Inner Solar System - Venus and Mercury as well as smaller solar orbits
- Science missions to the Outer planets
- Science missions leaving the solar system (with escape velocity from a 163,000 km altitude Apex Anchor)
- Alternative approaches to gaining extra velocity - trebuchets and or Slide release

One critical realization during this research was that with the lengthening of the space elevator and assembly of spacecraft at the Apex Anchor, the offer could be extended to all scientists to have:

- Any size/mass scientific equipment
- Any solar system destination
- Releases every day towards multiple scientific destinations
- inexpensively, routinely, safely, and environmentally friendly

Chapter 5 - Throughput Projection of Galactic Harbours

For Humanity to successfully and robustly move
off-planet, the question is:

What is the throughput of Galactic Harbours?

5.1 Introduction:

When designing large complex space systems, there is a process that develops
engineering projections of customer needs and then compares to infrastructure
capabilities. This chapter will show that Space Elevator and Galactic Harbour
development will follow historic approaches with an initial capability striving for
"what can be done" within the first few years and then building towards customer
requirements. Using the actual 2018 rocket launches as an initial capacity of lift-off to
orbit, and estimating the total throughput reaching orbit during that year, the results
were 114 launches with an average of eight metric tonnes yielding a total of about
1,000 metric tonnes during the year (rough estimate). In comparison, Elon Musk
estimates that he will need 1,000,000 metric tonnes[37] to support his Mars Colony - or
1,000 years at today's launch rate. During 2018, rockets went to various orbits (LEO,
MEO, GEO and interplanetary). These were robotic missions as well as human
missions, and had the flexibility to deliver all the customers' payloads to where they
wanted to go. This strength of going to specific orbits is why individual launches seem
so reasonable in cost and frequency. Rocket portals leverage changeable destinations,
optimized for LEO and MEO, and types of missions. In addition, they carry humans
safely and rapidly to their destinations in LEO.

However, we are now capable of developing a transportation infrastructure that can
satisfy future customer needs while doing it for less money, routinely (scheduled
daily), safely, in an environmentally friendly manner, and with massive movements of
payloads -- in other words, the Galactic Harbour strengths. This Galactic Harbour
transportation infrastructure, as shown in Figure 5.1, provides very strong support of
current and projected customer mission needs for GEO and beyond. If you think
about past transportation infrastructures, they tend to start out small (individual
point to point movement of cargo) while rapidly expanding towards the customers'
needs and wishes. First, infrastructures increase the capability of each individual
entity; and then, copy it until there are several transportation systems competing
within a given marketplace. This model of previous transportation infrastructure
growth will be used to show that increasing individual capabilities first, and then
multiplying the number of infrastructures, is probably the model that will be followed

---

[37] Musk, Elon., Quotation from CBS's Sunday Morning Show, 21 July 2019.

by Galactic Harbours. In the airline industry, for example, they first improved the airplane to haul cargo and people that then led to multiple airline companies competing for business. These historic infrastructures have shown that:

If you build it, and then improve it, they will come!

It is also extremely important to realize that the decade of the 2020's is going to be the initial period of movement off-planet. NASA, ESA, China, Russia and other countries are announcing their programs to go to the Moon and/or Mars. Russia has announced a plan to go to a Martian Moon and operate robots on the Martian surface as one of their first approaches to movement off-planet. The decade of the 30's will be remarkable with respect to interplanetary missions and cargo required to be delivered. This revolution across the space arena is active and aggressive. The new mass movement off-planet will add tremendous demands to our global lift-off capabilities.

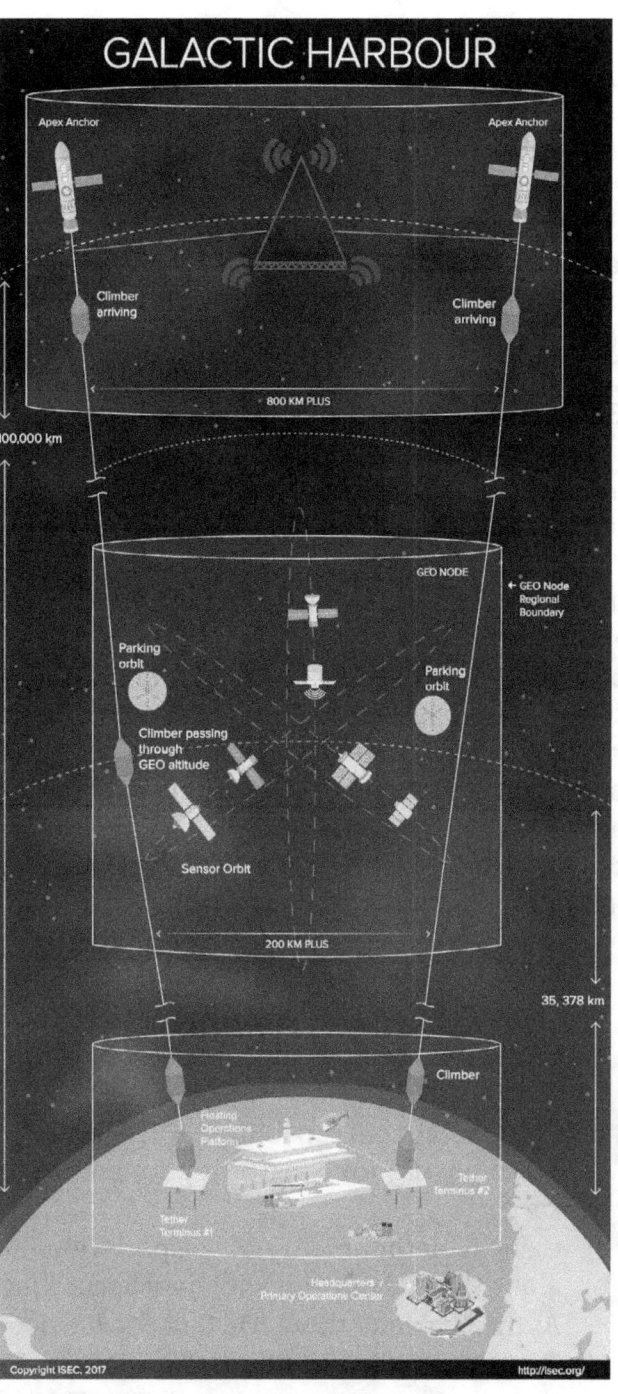

Figure 5.1 Galactic Harbour Concept

5.2 Growth of Galactic Harbours:

This chapter is looking to quantify the potential throughput of Space Elevator infrastructures around the world. Someone will build the first Space Elevator and then create a backup Space Elevator shortly thereafter so we do not ever become captured by gravity again. This pair of Space Elevators will become the transportation infrastructure for the Galactic Harbour. It is a combination of transportation infrastructure and business enterprises leveraging new capabilities. This is the creation of a capability (first and second Space Elevator) and then improving the capability over time. As the material of the tether grows stronger, mission successes will lead to increased customer demands. As such, this first Space Elevator will be created, duplicated, and then multiplied -- resulting in competing transportation infrastructures around the world. As described in Chapter Two's Vision, the number of Galactic Harbours will grow in numbers in response to needs and opportunities. As described before, within our time horizon, the initial one (Master Space Elevator) will prove the concept with a second Space Elevator creating a first Galactic Harbour. The competitive aspect of transportation infrastructures will lead to two and then three Galactic Harbours - first with Initial Operational Capabilities and then growing to Full Operational Capabilities. This vision for the 2035-2055 time period will be quantified and explained within the rest of this chapter. However, the authors believe this will only be the initial push for a robust capability to space with mid-century fostering growth towards six Galactic Harbours enabling robust movement off-planet.

During the Seattle ISEC International Space Elevator Conference in August of 2015, the International Academy of Astronautics study group #3-24 met. This group's product, at the end of four years with input from 47 global Space Elevator experts, resulted in a study report entitled "Road to the Space Elevator Era."[38] This included participates from Japan, USA, UK, Ukraine, France, Portugal, Russia, China, Canada and Brazil. The team agreed to use, as much as possible, consistent terminology for their report. This general list of terminology is shown in the Appendix G as a lexicon. The throughput of individual Space Elevators was estimated by the team creating common terminology. In the lexicon for Space Elevators, the following are in play for this study report, consistent with global understanding of the terms:

- Space Elevator (SE) - single tether of 100,000 km length
- Initial Operational Capability (IOC) - estimated capacity of 14 Metric tonnes of cargo each day.
- Full Operational capability (FOC) - estimated capacity of 79 Metric tonnes of cargo per day for a mature Space Elevator with human passengers as well as cargo.
- Galactic Harbour (GH) - Transportation Infrastructure with robust enterprises along the 100,000 km with dual Space Elevators.
- At the Horizon - three IOC Galactic Harbours
- Beyond the Horizon - three FOC Galactic Harbours growing to many around the equator.

When one emulates the growth of historic transportation infrastructures around the world, a parallel pattern comes into focus:

---

[38] Swan, P., David Raitt, John Knapman, Akira Tsuchida, Michael Fitzgerald, Yoji Ishikawa, Road to the Space Elevator Era, **Virginia Edition Publishing Company**, Science Deck (2019) ISBN-19: 978-0-9913370-3-3

- First a single Space Elevator
- Next the back-up Space Elevator becomes a robust partner
- Creation of a Galactic Harbour around the first two Space Elevators
- Next comes a second Galactic Harbour with one and then another Space Elevator.
- Followed by a third pair of Space Elevators inside another Galactic Harbour
- As engineering knowledge improves for tether materials and climber structures, throughput increases significantly towards an FOC capability.
- This transition of IOC to FOC throughput grows the transportation infrastructures towards very robust support for interplanetary missions.

When one looks at this century's maturation of the Space Elevator, there are many studies that have laid out the estimated carrying capacity of both the initial and more mature Galactic Harbours. This study depends upon previously established payload capabilities as documented in the next table. Some of the specific studies and papers that establish the projection of carrying capability are:

Table 5.1, Projection of Capacity and Schedule

Year	Sponsor	Study Title	Through-put Tonness and Start Date
2020	ISEC	Environmental Benefits from Space Elevators (in development)	14 in 2036 (IOC) & 79 in 2050 (FOC)
2020	ISEC	Interplanetary Mission Support (this report)	14 in 2036 (IOC) & 79 in 2050 (FOC)
2019	ISEC	Today's Space Elevator	14 in 2036 (IOC)
2019	IAA	Road to Space Elevator Era	14 in 2036 (IOC) & 79 in 2050 (FOC)
2014	IAA	Space Elevators: An Assessment of the Technological Feasibility and the Way Forward	14 in 2036 (IOC)
2013	Obayashi	The Space Elevator Construction Concept	79 in 2050 (FOC)

IAA - International Academy of Astronautics, ISEC - International Space Elevator Consortium, Obayashi - The Obayashi Corporation

Taking that flow of development, one would have something similar to the following:
- Operations Date (IOCs)
  - 2037 - First IOC operational tether
  - 2038 - Second tether becoming a first Galactic Harbour
  - 2040 - Second Galactic Harbour
  - 2041 - Third Galactic Harbour
- Operations Dates (FOCs)

- o 2047 - First Galactic Harbour with one FOC Space Elevator and one IOC Space Elevator
- o 2048 - First Galactic Harbour with two FOC Space Elevators
- o 2050 - Second Galactic Harbour with two FOC Space Elevators
- o 2051 - Third Galactic Harbour with two FOC Space Elevators

In addition to the definition of terms (SE, GH, IOC, FOC), the estimated schedule has been projected. This was first shown in the ISEC 2019 Study Report (Today's Space Elevator). This schedule estimates the growth of capability and schedule as described in the last series of discussions. This projected schedule shows major steps in the development of this new transportation infrastructure called Galactic Harbours.

Table 5.2, Space Elevator Possible Schedule[39]

Event Occurring between two dates	Early	Estimated
Material for Tether shows required Characteristics	2020	2022
Material developed for Space Elevator Tether	2023	2029
Major Segments Validation Testing	2024	2030
Integrated Orbital Testing (Low Earth Orbit)	2031	2032
Launch of Deployment Satellite	2032	2034
Deployment of Space Elevator	2033	2036
Buildup of Space Elevator to Initial Operations Capability	2037	2040
Initial Operations	2037	2040
Second Space Elevator operational	2038	2041
Galactic Harbour Operational (IOC capacity)	2037	2040
Second Galactic Harbour Operations (IOC capacity)	2039	2042
Third Galactic Harbour Operations (IOC capacity)	2040	2043
First Galactic Harbour with Full Operations Capability (with People)	2047	2057
Second Galactic Harbour with Full Operations Capability (with People)	2048	2057
Third Galactic Harbour with Full Operations Capability (with People)	2049	2057

5.3 Throughput of Space Elevators:

After this recognition of previously established carrying capacity and operational dates, capability grows within the transportation infrastructure as shown by the next

---

[39] Swan, Peter, Michael Fitzgerald, "Today's Space Elevator," ISEC Study Report, lulu.com, 2019.

chart. This development from a single IOC Space Elevator to three Galactic Harbours with the full capability estimated to handle humans and cargo illustrates the remarkable revolution in lift-off capability to support interplanetary missions to multiple destinations. The increase in capability over time is shown in the next chart.

Figure 5.2, Galactic Harbour Throughput (Metric Tonnes)

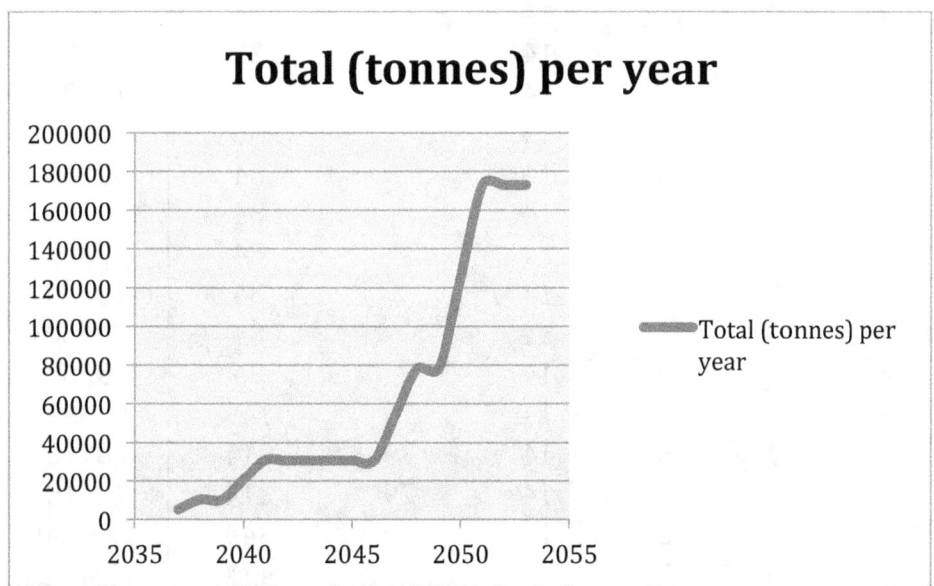

The growth is amazing when one thinks of the limited capability of current rockets (one year's launch capability is less than 1,000 Metric tonnes) with the first year of a single Space Elevator operations giving 5110 metric tonnes /year to GEO and beyond. This rapidly grows to six times that for three Galactic Harbours each with two IOC Space Elevators (30,660 metric tonnes /year). However, when one grows to an FOC capability in each Space Elevator for three Galactic Harbours, the numbers are remarkable and mission enabling (173,010 metric tonnes /year).

When one looks at the table that created the previous chart, it looks like this:

Table 5.3 Total Throughput by Year

Year	# IOC SEs	# FOC SEs	# GHs	Tonnes per IOC	Tonnes per FOC	Total (Tonnes) per day	Total (Tonnes) per year
2036	0	0	0				
2037	1	0	1	14		14	5110
2038	2	0	1	14		28	10220
2039	2	0	1	14		28	10220
2040	4	0	2	14		56	20440
2041	6	0	3	14		84	30660
2042	6	0	3	14		84	30660
2043	6	0	3	14		84	30660
2044	6	0	3	14		84	30660
2045	6	0	3	14		84	30660
2046	6	0	3	14		84	30660
2047	5	1	3	14	79	149	54385
2048	4	2	3	14	79	214	78110
2049	4	2	3	14	79	214	78110
2050	2	4	3	14	79	344	125560
2051	0	6	3		79	474	173010
2052	0	6	3		79	474	173010
2053	0	6	3		79	474	173010

5.4 Comparison of Mass Throughput:

The demand for supplies and support equipment will be remarkable as off-planet colonies are developed. There are several benefits from Galactic Harbour transportation infrastructures: first, there is the tremendous load capability; second, remarkably fast transits are available; and third, departure dates towards interplanetary destinations can be every single day of the year.

Massive Movement to Mars   Movement of people and mission payloads by rockets has been inspiring with trips to the Moon and the delivery of constellations of communications and navigation satellites. However, all these successes have leveraged burning rocket fuel, limited by the rocket equation. Defeating gravity costs enormous amounts of energy with the reality being that delivery to mission orbit consumes both fuel and often the vehicle. A quick comparison of rockets to Galactic Harbours should explain the strengths of a permanent infrastructure for access to space. Figure 5.3 shows an estimate for delivery of payload mass to orbits when using the rocket equation. The average mass that gets to Low Earth Orbit from a rocket is only 4% of the mass that started on the pad. The average to GEO is only 2% of the pad mass while getting to the surface of the Moon (LS) is only 0.5% of launch pad mass. Leaving by way of the trans-lunar injection (TLI) results in only 1.5% of pad mass (similar percentage to go to Mars). To deliver a one kg pizza to the Moon's surface (assume includes cheese), the mass at the pad would have been 200 kg. The strength of Galactic Harbours is that they are a permanent infrastructure delivering supplies to GEO and beyond. If you wanted to deliver a one kg pizza to the Moon thru the infrastructure, you would put one kg in the tether climber cargo bay.

Figure 5.3, Percentage of mass at Destination (rocket equation results)

0.5% to LS

1.5% to TLI

2% to GEO

4% to LEO

Structures, motors, Fuel = 96%

Notes: Average #s, Many variables, re-useable saves $

In addition to delivery benefits (factor of 200 to one), Space Elevators also have a capability of adding energy to the payload by raising it (potential energy) and then releasing it at high velocity (rotational energy from the planet). This increase in energy at departure from the Apex Anchor allows payloads to reach their destinations more rapidly with departures possible every day of the year (some days have quicker pathways than others depending on destination and planetary alignment). In addition to the efficiency of permanent infrastructures, departure opportunities are far more frequent - ensuring movement of more supplies and equipment. This is shown with delivery of 14 metric tonnes each day by a Space Elevator x 3 Galactic Harbours (x2 tethers per) or 84 metric tonnes per day. When this grows to 79 metric tonnes each day by each Space Elevator as a mature Full Operational Capability, the throughput becomes 474 metric tonnes daily.

Table 5.5: Number of Rocket Launches vs. Galactic Harbour Lift-Offs

Type	Lift Average (Metric tonnes)	Launches per year	Total Mass to Interplanetary (Tonnes) per year
Individual Heavy Launch Vehicles	10 per launch	87 (average last 5 years)	Less than 1,000 metric tonnes to single missions
Future Vehicles (SpaceX's Starship)	100 per launch reusable - requires refueling (x3 to go to Mars - x2 for Moon)	100 StarShips + 300 tankers (guess)	10,000 metric tonnes towards Mars
Galactic Harbour Transportation Infrastructure at IOC	6 tethers x 14 tonnes = 84 tonnes per day	every day towards Mars and Moon	84 x 365 = 30,000 metric tonnes
Galactic Harbour Transportation Infrastructure at FOC	6 tethers x 79 tonnes = 474 tonnes per day	every day towards Mars and Moon	474 x 365 = 173,010 metric tonnes

This table shows a comparison of today's average rocket capability per year (of 1,000 metric tonnes to GEO and beyond) with Galactic Harbours' 30,660 metric tonnes per year. An estimate for future rocket capability has been shown (5,000 metric tonnes /yr). The comparison uses total capability to lift off the ground towards the Moon and Mars - with no other misssions. In addition, the launch numbers are for current launchers (the future with more robust reusability and perhaps even larger capacities with Blue Origin's New Glenn and SpaceX's StarShip should increase throughput). Even with these future rocket performance numbers, the difference is still tremendous, with advantage Space Elevators. This monumental advantage is because Galactic Harbours are a permanent transportation infrastructure with no wastage from the extravagant rocket equation.

The next chart shows the comparison of these capabilities to orbit with three customer demands of the future. These Reference Destinations, as described with stated needs in Chapter 3, are compared to current and future capabilities of rocket portals and Space Elevator infrastructures. The chart is calculated as if each liftoff capability is dedicated to each of those reference missions; however, everyone knows there will be sharing of capacity across all missions. One obvious conclusion is that the sharing of delivery of cargo should cross both transportation portals and infrastructures, with rockets focusing on historic orbits and human flights while Galactic Harbours focus on GEO and beyond. As this chart shows, future needs exceed rocket portal capabilities by orders of magnitude and even stretch the capabilities of Galactic Harbours. The good news is the duration of lift-offs for these missions supports the developmental times of mega-projects across decades.

Table 5.7, Destination vs. Years to Satisfy

Destination Needs vs. Liftoff Capacity	Capacity	Projected Rate Historic Missions to 2040	Destination Needs - Mars Colony	Destination Needs - Moon Village	Destination Needs - Space Solar Power
		Metric Tons	Metric Tons	Metric Tons	Metric Tons
Needs by 2040 (Metric Tons)		5000	1000000*	500000**	5000000***
	MT/yr	Years to Satisfy	Years to Satisfy	Years to Satisfy	Years to Satisfy
2019 Rockets to Orbit	1,000	5	-	-	-
Rockets for 2040^	6500	0.8	154	77	770
Initial Space Elevator (3036)	5110	1	200	100	1000
3 GHs (6 IOC SE) (2040)	30660	0.016	32.6	17	150
3 GH (6 FOC SE) (2052)	173000	-	5.78	2.89	28.9

^ Rockets for 2040 estimated at 50 Starships (100 MT) + 150 old (10 MT) per yr   * Musk Estimate   ** Team Esitmate   *** Dr. Mankins Estimate

5.5 Conclusions:

This throughput chapter showed that the potential movement of mass off-planet by Galactic Harbours will enable the achievement of major missions hindered by the limited capabilities of the past. This transportation infrastructure will satisfy customer needs while being compatible and complementary to the growing rocket portals. Each will have strengths to support various customers; however, the movement of cargo for complex and massive undertakings is a natural strength of the Galactic Harbours. Complementary transportation portals and infrastructures can ensure success in the different missions and destinations desired by future movement off-planet.

Table 5.7, Galactic Harbour Fulfillment of Reference Missions

Reference Mission	Metric tonnes to Destination	Galactic Harbour IOC Fulfillment Time (yrs)	Galactic Harbour FOC Fulfillment Time (yrs)
Space Solar Power	5,000,000	150	29
Mars Colony	1,000,000	33	6
Moon Village	500,000 estimated	17	3

Leveraging a previous chart from Chapter 3, Mission Needs, the delivery times for the three Reference Missions are shown. This tends to put the whole picture into focus. The demands are huge for these critical reference missions and their destinations. FOC Galactic Harbours are needed as soon as possible to support humanity's dreams.

A Galactic Harbour Infrastructure Network, with a complementary and compatible set of rocket portals, will Enable Humanity's bold future off-planet.

Chapter 6 - Galactic Logistics Supply Chain

6.1 Introduction:

In the broad field of Transportation Systems, the 19th and 20th Centuries saw evolutionary advances in the movement of cargo and people across vast distances. In the United States, the opening of the Transcontinental Railroad in the 1860s brought expansion and opportunity while making such systems as covered wagons, pony express and even "around the Horn" sailings obsolete. The basic elements of today's global cargo transportation system include ocean going ships, railroads, trucks and cargo planes. As a generalization, the most cost-effective way of transporting large quantities of cargo over long distances has been, and will continue to be, by ocean going vessel. On the other hand, small quantities of time sensitive, high value cargo are typically transported by commercial airlines. In the 1970's the concept of international standard shipping containers became a reality and has developed, in the ensuing years, as the primary mode of moving cargo around the globe. *Containerization* led to *Intermodalism*. Under this now common system, a product can be secured inside a container at its origin, then trucked to a regional seaport, placed aboard a container ship (along with thousands of others), shipped across an ocean to another seaport, placed upon a specialized "unit" train, transported across country to a distribution center in a regional city and finally removed from the original container and placed onto a local truck taking it to its ultimate destination or market.

The various entities and businesses involved in this transportation system form what is known as the global logistics supply chain. These entities include shipping lines (ship owners), seaport and rail terminal operators (e.g. stevedoring firms), railroad owners and operators, government customs officials, customs agents, freight forwarders, trucking companies, labor unions, public port authorities, etc. In the transportation industry, there is a large body of technical knowledge that is commonly called supply chain management. This body of knowledge is becoming increasingly sophisticated by using advances in computer and communications technologies, including blockchain technology and the so-called Internet of Things (IOT) to keep track of shipments from source to destination. Not just the containers themselves, but their contents. These systems rely on remote sensing technology and digital interconnectivity on a global scale. With all its complexity and world-wide scope, today's transportation system is basically two-dimensional. The movement of cargo (and people) is either by land, on the water or by air, well within the Earth's atmosphere, or by some combination thereof. The present global logistics supply chain, robust as it is, must adapt and expand to handle the needs of future geosynchronous orbit and interplanetary support (including the Moon). But how do we do this? Today, it is moved to a rocket launch facility somewhere on land or on the ocean using existing transportation systems, then launched into space using conventional government owned or private booster rockets. Even today, each lift off and return are not routine; they are "events."

The Elevator is no longer a mystery. Engineering approaches for the Tether, Tether Climber, Earth Port, GEO Region, and Apex Anchor have been expressed in terms everyone understands: a harbor. The last technology hurdle – strong material for the Tether – has been conquered. The Galactic Harbour is a triad of change: a transportation force of the future; an enabler of robust space-based enterprise; and the initial infrastructure of the 3rd dimension of Earth's transportation and logistics system. The Galactic Harbour will indeed become where transportation and business will meet

6.2 Background:

ISEC's Tech Ready assessment is accepted by many experienced in "megaprojects" as a starting signal: for space megaprojects[40]. Efforts should be taken to start engineering validation - a test heavy time when engineers determine how to use technologies in the build process. The Tech Ready assessment is a signal to industry that a Space Elevator can be built. Those companies conducting engineering validation testing will be seeking Space Elevator performance requirements - levied by the coming interplanetary missions. This Tech Ready statement is made because the Space Elevator community must seek collaboration with the transportation community. The Galactic Harbour architects are starting to develop the requirements baseline for this new transportation infrastructure. This chapter will help initiate discussions with other communities seen as essential to Galactic Harbour successes. This collaboration begins with an amplified discussion of how Galactic Harbours operate, and especially how they will support the three leading space destinations of our time: Enterprises at GEO, missions to Mars and Moon (as explained in Chapter 3, GEO and Interplanetary Mission Needs).

With several missions foreseen, several Harbours are needed to meet massive cumulative throughput requirements. (Throughput was discussed and quantified in Chapter 5) Additionally, the sources of mission support cargo will be diverse - likely world-wide. To keep the cargo flowing, the supply chain's deliveries are best optimized by a multi-path flows within multiple Harbours. Each Harbour is a member of the worldwide network of Galactic Harbours, all are located on or near the equator around the world; and, all are delivering payloads to space while supporting mission operations.

By the middle of the 21st Century the first of several Galactic Harbours will provide a nexus for connecting Earth's two dimensional transportation systems of rail, highway, shipping, and air to the third dimension providing a safe, reliable and cost effective

---

[40] Swan, Peter, Michael Fitzgerald, "Today's Space Elevator," ISEC Study Report, lulu.com, 2019.

means to move material into orbit and beyond. This "material" being the logistics underpinning of a colony on Mars, processing facilities on the Moon, and the many enterprises populating the GEO Regions. The Elevators in each Harbour will not be exactly the same. It seems evident that a Galactic Harbour focused on supporting a variety of enterprises at GEO will be different than one solely supporting interplanetary travel to and from the Apex. One Harbour here will complement another Harbour there.

The Climber that supports enterprises at GEO may have distinct differences from the Climber that delivers cargo to the Apex Anchor ready for travel to Mars or the Moon. Another aspect of the operational Harbour network is the manner in which they support each other. An enterprise at one GEO Region may be involved with providing products destined to Mars on a cargo craft leaving from an Apex Region atop another Galactic Harbour.

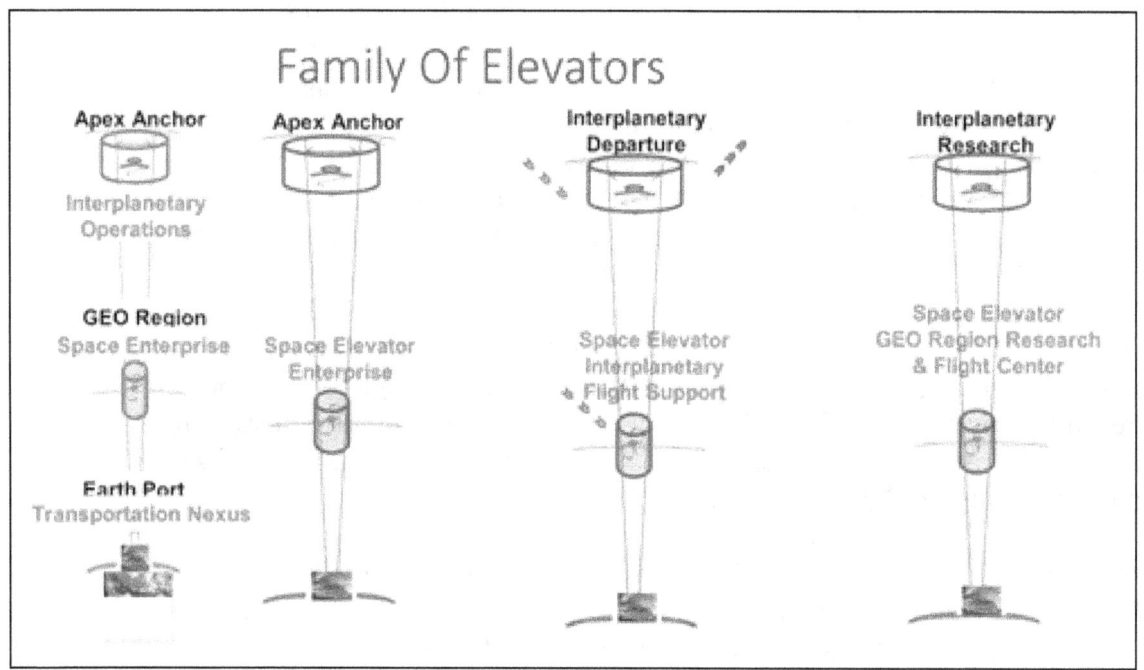

Figure 6.1, Family of Galactic Harbours

6.3 Earth's Galactic Harbours, a Vertical Transportation System:

As mentioned earlier, international standard shipping containers are the primary mode of moving cargo around Planet Earth. *Intermodalism* is now the common system. A product can be secured inside a container at its origin, tracked by an Internet of Things (IoT) standard, and arrive at a destination in the same container. The Galactic Harbour's Earth Port will be the entry point for the vertical off-Earth travel of that product. The product will move from mode to mode, with repackaging of the product or its container only where necessary to be compatible with space travel and use at its in-space destination. The Galactic Harbour is the core capability of the coming space transportation infrastructure and will adhere to the intermodal transportation standard. Galactic Harbours will be seen as familiar, robust amalgams of transportation and enterprise and will attract users, investors, partners, entrepreneurs, and supporters. In the near term, new specialized Galactic Harbours will operate - expanding Earth's transportation infrastructure to access space and melding with the entire, diverse, mosaic of space.

6.4 Transportation Infrastructure Matures:

At the horizon, the first Galactic Harbour will initially support nascent enterprise activities along the GEO belt. Soon afterwards, factories and solar power generation systems will be assembled near GEO and efficient interplanetary departures from the Apex will begin. Some earthbound products and materials from GEO will arrive at the Earth Port. The Earth Port becomes a valued nexus of transshipment and trade. On orbit, businesses will flourish, satellites will be repaired and refueled, new products developed, spacecraft and facilities constructed, solar power collection systems will be assembled, and interplanetary journeys will be launched to a variety of destinations. The world's transportation infrastructure will be connected from Main Street to the Solar System via Galactic Harbours.

6.4.1 Insight into GEO Region: The importance of the Galactic Harbour's GEO Region to both enterprise development and interplanetary travel cannot be overstated. In fact, the GEO Region begins playing a major role early in the overall deployment of the other segments of the Harbour. The GEO Region will be active early on because a large number of specialized support spacecraft will be parked there, starting as early as Single String Testing (Sequence Step #3). The support spacecraft will execute, monitor, and record the extensive testing and retesting process through the sequences approaching first operations.[41]

When a Space Elevator becomes operational, it will service the enterprises within the Harbour. These enterprises will be tenants of the Harbour following the relationship model seen in most major harbors around Planet Earth. The Space Elevator Transportation System will operate the Harbour, as a good portion of its activities will be supported by enterprises within the Harbour and located in the GEO Region. As the Galactic Harbour approaches first operations, a number of "public utility" type enterprises will be established: power, safety, some types of communications and computing services, and the like. The first independent enterprises within the Space Elevator Enterprise System will be pre-existing on-orbit businesses relocating to the GEO Region. Satellite refurbishment services, such as on-orbit refueling and repair, are expected to start along the GEO belt by the late 2020's. These nascent enterprises

---

[41] Fitzgerald, Michael, Architectural Note #6, 7, 8: Sequences, www.isec.org Architect Page

would see the advantage of moving to the Galactic Harbour for better supply chain access. The second generation of enterprises will arrive at GEO soon after the Space Elevator Transportation System begins daily operations. With the Galactic Harbour's main channel open and flowing, small spacecraft assembly and related assurance flight testing will follow. A Space Tug service company will arrive for intra-Region transportation; supporting payloads delivery from Climber transshipment point to GEO Region clients. The expected Space Based Solar Power assembly will follow quickly: first with gathering the needed "ready to assemble" parts brought by Climbers and then very active periods of assemble – test – assemble - test until the modular power collection enterprise opens for business.

6.4.2 Insight into Apex Anchor:

Concurrent with the activity at GEO, the Apex begins to build flight operations enterprise facilities for a variety of operations. Flights will arrive and depart not unlike flight operations at a major airport today. Some pre-assembly for these Apex structures will be accomplished at GEO, but unlike facilities at GEO, facilities at the APEX Anchor must remain connected to the Tether. The Apex will have a number of transshipment fixtures able to handle cargo coming up the Tether and preparing loading onto logistics craft headed to the Moon, Mars, the Asteroid Belt, etc. The Apex Anchor also has a fundamental role in the operations of the Galactic Harbour. It is the classic counterweight for each Space Elevator with each Harbour. As such, it must maintain mass awareness and remain in coordination with the HQ/POC in that regard. Any arriving or departing Climber, and any arriving or departing cargo craft will change the overall mass at the Apex Anchor. In the future, the Apex will have the ability to extend beyond its nominal altitude to adjust the departure and / or landing velocity of interplanetary flights. The concept baseline of the Galactic Harbour has a Tether Reel In / Reel Out (RIRO) mechanism at the Apex Anchor. It is used to keep the tether taut no matter the mass. A duplicate RIRO mechanism will play a key role in future flight velocity management operation. In addition, the Apex Anchor is a key player in the control of the dynamics along the tether.

Massive computing power is needed to manage the Apex Anchor, generate flight plans for the daily mission support departures, control incoming arrivals from across the solar system, and numerous other tasks in "Outer Space." The Apex will be robotically preparing Flight Systems for departure. That preparation could include some detail final assembly, perhaps attaching the selected specialized thruster used for the cargo craft's velocity adjustments along its route. Other departure and arrival adjustments will be needed. As will precision handling of cargo & material on both departing and arriving flights. It is best that a container not be dropped - or a wrench for that matter. It will move away quickly. The computing power load will also be needed to handle a range of surveillance functions. The Apex Anchor is at the edge of Earth's domain. Sensors to look outward will be of great value. The Apex is also the "high ground" offering grand synoptic views of a large portion of activities between the Apex and Planet Earth below.

6.4.3 Insight into Earth Port - POC: On the first day of operations as a Galactic Harbour, the Earth Port will already be in its second decade of operations. The Earth Port's Floating Operations Platform (FOP) will have supported a wide variety of test and operation rehearsal activities over the years. The FOP's top deck holds the various communications systems necessary for Harbour operations - managed by the Galactic Harbour's Primary Operations Center (POC). Both the top deck communications and the operations center will be essential for supporting the robust taxonomy of testing needed to carry the Space Elevator Transportation System development leading to the Limited Operations Phase and then Initial Operations. The POC will be the heart of these operations, and it will be busy. Simply stated the POC must keep track of every object within the Galactic Harbour, and know where every object is heading; keeping all of them under positive control and awareness. Customers and clients will need to know when their payload arrives at its destination so that they can secure it and put it to immediate use. In addition, Galactic Harbour operators will need full awareness of any nearby object.

The Galactic Harbour's Primary Operations Center will manage the entire Harbour via a fully networked command & control structure. The operating mantra of that structure will be "live, virtual, and constructive." Objects within the Harbour will either be attached to, in orbit around, coming into, or leaving the Harbour. Objects will either be cooperative with the POC, or not. All cooperative objects will repeatedly report their position and (if in orbit) their heading. Non-cooperative orbiting objects will be treated as a "threat" until they are tagged or depart the Harbour. Cargo items aboard Tether Climbers will have their own version of the RFI/IFF tag. Until the item gets to its Harbour destination it will be tracked much like any delivery service does today.

Figure 6.2, Earth Port Floating Operations Platform (GHA image)

The Galactic Harbour's POC will operate as a Space Command & Control Center. It must view its entire domain. It will use a range of sophisticated sensing systems and information analysis tools - artificial intelligence will play a huge role. The POC must account for arriving and departing items whether cooperative or non-cooperative. It must ensure that cargo gets to where it is going and that the Harbour operates safely and efficiently. The POC will operate a "hot backup" in the Access City for safety and continuity of operations purposes. In truth, an additional hot backup may be operated at a secure location to ensure that a consensus decision can always be achieved.[42]

6.4.4 Insight into Earth Port - Transshipment: Material, supplies, and cargo arrive at the Earth Port's Floating Operations Platform (FOP) heading to multiple destinations. Some will be dropped off for the FOP itself; some will head to the GEO Region for Galactic Harbour operations or the several Enterprises operating there; and significant numbers will be headed to the APEX Anchor for the flight operation activities there or onto a supply craft for interplanetary travel. All of the cargo arriving at the FOP has been preliminarily sorted by destination and source certified at the Supply Chain Management Center in the Access City. Payloads headed to space maybe, at least to some degree, repackaged for that. Thermal and radiation protection are essential, and each container will have a range of monitoring devices attached. The packaging / re-packaging locations will be agreed upon between the payload providers and the Space Elevator operator. The FOP has several "clean rooms" that will support transshipment functions such as final placement of the payload onto the Climber pallet and then integrating that pallet into the Tether Climber. The last Earth Port transshipment function is release of the Tether Climber onto the Tether for its journey into space.

6.4.5 Insight into Earth Port - Access City: The Earth Port's Access City is home to the Galactic Harbour's headquarters and the online backup(s) for the POC. The Access City is also home to the Harbour's Business Operations Center. The Supply Chain Manager for all cargo going into the Harbour and up the main channel will be monitoring from there. Items arrive from manufacturers at the Access City and are certified for travel and assigned to a Climber - soon departing up the Tether. Access City efforts include acceptance procedures to ensure that the Galactic Harbour is receiving material from certified sources. Two cargo types are evident. Material designated for use within the Space Elevator Transportation System (e.g. Elevator spare parts and consumables) and client's payload.

The operations of the Space Elevator, both as a transportation infrastructure and a business enterprise, must seamlessly mesh with existing and future global transportation system. The physical point (nexus) where this meshing of systems should logically occur would begin at those HQ/POC facilities located at or near the

---

[42] Note from Harbour Master: "The components of the Earth Port segment of the **first** Galactic Harbour will be designed to operate on a 24 hour basis year round. The anticipated **daily** payload capacity of the tether climbers at IOC is **14 MT.** This throughput capability may vary from day to day due to operational realities at the Earth Port location. However, the aggregated yearly payload delivery projections used in this paper are valid for comparative purposes."

Earth Port Access City. Everything headed into and/or through the Galactic Harbour arrives at Supply Chain Management Center in the Access City.

6.5   Conclusions:

The magnitude of this portrayal is humbling. This portrayal of a new capability (vertical transshipment to space) has a destination well-marked by the needs of humanity. It was seen by those who preceded us and it will be built by those who follow.

- In the last few years, ISEC has taken a simple notion and defined a uniquely capable Space Elevator based Galactic Harbour. This multi - year effort led to the Tech Ready announcement in 2019. Such efforts often were the prelude of mega-project efforts called "technology push." The Space Elevator community now turns to an effort to examine the purposes of the Galactic Harbour.
- At the horizon, the Earth Port becomes a valued nexus of trade. Galactic Harbours support a multitude of business activities in the GEO Regions. The Apex Anchor is the base for efficient interplanetary departures, and Galactic Harbours become the foundation for reaching out to our solar system.
- Beyond the horizon, several Galactic Harbours will operate; expanding Earth's infrastructure to access space and melding it with the entire, diverse, mosaic of space.

The transportation story of the 21 st Century is still being written.
The Apex is where the Galactic Harbour meets Outer Space and
this Story meets the "Last Frontier."

Chapter 7 - Conclusions

7.1 Introduction:

The Galactic Harbour is the unification of the Space Elevator Transportation system and the commercial Enterprise ecosystem. As humanity expands off of Earth, the need for support increases exponentially. The mass per day required to be delivered to the Moon, Mars and other destinations will stress out the current rocket based approach -- something revolutionary is required. The authors project into the future and see that Space Elevators will allow the robust growth of humans off-planet to accelerate with three major improvements: massive movement of mission support equipment, a tremendous opening up of launch windows, and shorter travel times for interplanetary and lunar journeys.

This report recognized and expanded upon strengths of the Galactic Harbour not exploited before. They are shown in Figure 7.1, and summarized as:

- Fast transit to destination (Mars as short as 61 days). Arizona State University (ASU) research showed that release from an Apex Anchor enables remarkable transit times during the 26 month repeating orbital relationship between Earth and Mars.
- Massive liftoff capability (an initial capability of 14 metric tonnes payload per day). Space Elevators start out with huge throughput capacity with daily liftoffs (5,110 metric tonnes per year). In addition, there will be remarkable growth as the tether material and infrastructures mature to Full Operational Capability reaching 79 metric tonnes per day (178,000 metric tonnes per year).
- Daily departures available (no waiting for 26 month Mars launch windows). The ability to launch each day towards Mars is a revolutionary concept vs. the traditional wait period of 26 months.

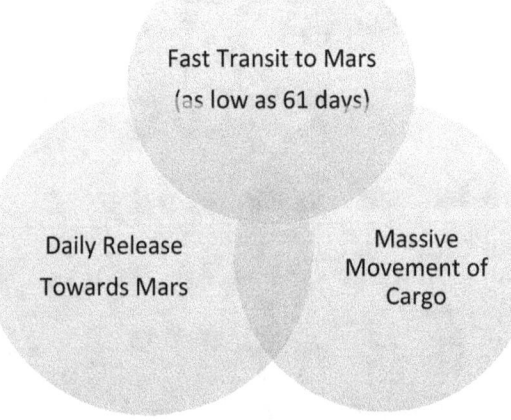

Figure 7.1, Newly Recognized Galactic Harbour Strengths

Fast Transit to Mars (as low as 61 days)

Daily Release Towards Mars

Massive Movement of Cargo

7.2 Developmental Progress:

A "Sea State Change" has occurred within the developmental progress of Space Elevator concepts. Significant activities have occurred worldwide, as reflected in the International Academy of Astronautics (IAA) 2019 study report "A Road to the Space Elevator Era."[43] These engineering activities have led to a successful Preliminary Technological Assessment. This surfaced after multiple organizations accomplished major tasks:

- ISEC produced ten year-long studies with resulting reports (Appendix D).
- The IAA produced two study reports supporting the concept (Appendix E).
- The Obayashi Corporation conducted an independent study that focused upon humans on the Space Elevator and massive movement of Space Solar Power satellites to GEO.[44]
- Internal ISEC assessments were provided by a series of Chief Architect's Notes. (see www.isec.org).
- The agendas of major international space agencies and private companies are aligning to target human presence and/or settlements on the Moon and Mars, thereby establishing demand pull.

Because of these achievements, many questions have been answered over the last ten years with respect to the readiness of Space Elevators. The position of ISEC is that the:

> *Space Elevator development has gone beyond a preliminary*
> *technology readiness assessment and is ready to enter initial*
> *engineering validation testing -- leading to the establishment of*
> *needed capabilities.*[45]

7.3 Galactic Harbour Expands the Global Logistics Chain:

With all its complexity and world-wide scope, today's transportation system is basically two-dimensional. The movement of cargo (and people) is either by land, on the water or by air, well within the Earth's atmosphere, or by some combination thereof. The present, global logistics supply chain, robust as it is, must expand to handle the needs of future geosynchronous orbit and interplanetary support missions (including the Moon). But how do we do this? Today, it is moved to a rocket launch facility somewhere on land or on the ocean using existing transportation systems,

---

[43] Swan, P., David Raitt, John Knapman, Akira Tsuchida, Michael Fitzgerald, Yoji Ishikawa, Road to the Space Elevator Era, **Virginia Edition Publishing Company**, Science Deck (2019) ISBN-19: 978-0-9913370-3-3

[44] Ishikawa, Yoji, The Space Elevator Construction Concept, Obayashi Corporation, 2013, IAC-13-D4.3.6.

[45] Swan, Peter, Michael Fitzgerald, "Today's Space Elevator," ISEC Study Report, lulu.com, 2019.

then launched into space using conventional government owned or private booster rockets. Even today, each lift off and return are not routine; they are "events." Over the last few years, these events totaled approximately 100 per year across all nations. The first Space Elevator will be scheduling for 365 liftoffs per year.

By the middle of the 21st Century the first of several Galactic Harbours will provide a nexus for connecting Earth's two dimensional transportation systems of rail, highway, shipping, and air to the third dimension providing a safe, reliable, Earth friendly, and cost effective means to move material into orbit and beyond. This "material" is the logistics underpinning of a colony on Mars, processing facilities on the Moon, and the many enterprises populating GEO Regions. The Space Elevators in each Harbour will not be exactly the same. It seems evident that a Galactic Harbour focused on supporting a variety of enterprises at GEO will be different than one solely supporting interplanetary travel to and from the Apex Anchor.

7.4 Arizona State University Research:

Associate Professor Matthew Peet and Dr. Peter Swan of ISEC initiated research in the spring of 2019 assessing the strengths of releasing payloads at the Apex Anchor for interplanetary missions. The students, led by James Torla, examined the potential impact of a Space Elevator Apex Anchor release of cargo for permanent human habitation on Mars and the Moons of Jupiter and Saturn.[46] The question, when looking at it from a Space Elevator perspective, became "can the Galactic Harbour infrastructure do daily launches with a variety of flight times and distances to Mars?" The research leading up to this study report developed major conclusions that will illustrate the remarkable strengths of Galactic Harbours. A quick summary of the strengths identified during the research, and illustrated in this report, are:

- Release from the Apex Anchor opens up daily trips towards the Moon and Mars.
- Release from Apex Anchor enables rapid transit to the Moon (14 hours) and Mars (as short as 61 days).
- Multiple Galactic Harbours are shown as the transportation vision of the future supporting significant missions at GEO and beyond.
- Massive movement of payload cargo enables these missions.
- Routine (daily), safe, environmentally friendly, inexpensive, with continuous massive movement of cargo will enable movement off-planet.
- Movement off-planet will require complementary infrastructures, such as rockets and Galactic Harbours, each with their own strengths and short-falls.
- Permanent Space Elevator Transportation Infrastructures will move massive amounts of cargo for support of off-planet missions.

---

[46] Torla, James and Matthew Peet, "OPTIMIZATION OF LOW FUEL AND TIME-CRITICAL INTERPLANETARY TRANSFERS USING SPACE ELEVATOR APEX ANCHOR RELEASE: MARS, JUPITER AND SATURN," International Astronautics Congress (IAC-18-D4.3.4), Washington D.C., 2019.

- The engineering challenges and tether material development are expected to match the schedules shown in this study report (also, see IAA study "Road to the Space Elevator Era"[47]).

Overall conclusions from the ASU research:

- Interplanetary travel is greatly enhanced by release from the Apex Anchor in time of flight, required delta velocity and launch opportunities.
- Release velocities at the Apex Anchor expand freedom of launch for both scheduled and emergency responses (as short as 61 days to Mars and 14 hours to the Moon).
- Low cost options exist when looking at the lower delta velocity opportunities
- Specialized missions can be designed based upon predictable orbital positions and the rotation of the Apex Anchor around the Earth. Low time of flight missions are especially valuable to mission directors, especially in emergency situations.

7.5 Reference Missions:

During this study, three destinations were chosen as reference missions: Space Solar Power, Mars Colony and Moon Village. When one looks at the tremendous demands to support the customers, it is obvious that rockets alone will limit missions. With cooperative activities tying rocket portals to Galactic Harbour infrastructures, the reference missions seem possible in the desired times. This is shown in the Chapter 5 chart: Total Throughput. Specific needs for individual missions can be assessed with some estimates already expressed. To place this whole study in perspective, the comparison of "demand pull" for these three Reference Missions is identified as:
- GEO Base - Space Solar Power - 5,000,000 metric tonnes
- Moon Base - Lunar Village - 500,000 tonnes
- Mars Base - SpaceX Colony - 1,000,000 metric tonnes

7.6 Throughput by Space Elevators:

After discussing carrying capacity and operational dates, projections of capability growth within the global transportation infrastructure is shown by the next chart. This development from a single IOC Space Elevator to three Galactic Harbours with the full capacity estimated to handle humans and cargo illustrates the remarkable revolution in lift-off capability. The increase in this capacity over time is shown in Figure 7.2.

---

[47] Swan, P., David Raitt, John Knapman, Akira Tsuchida, Michael Fitzgerald, Yoji Ishikawa, Road to the Space Elevator Era, **Virginia Edition Publishing Company**, Science Deck (2019) ISBN-19: 978-0-9913370-3-3

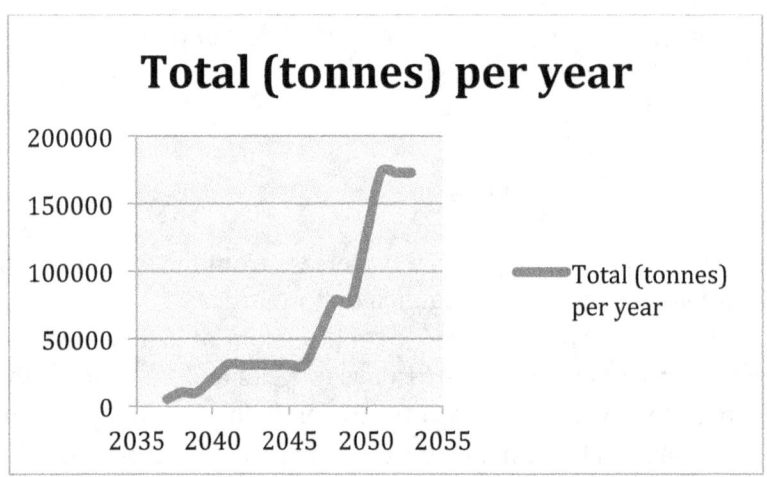

Figure 7.2, Galactic Harbour Throughput

This study report shows that the potential movement of mass off-planet by Galactic Harbours will enable the achievement of major missions that were hindered by the limited capabilities of the past. This transportation infrastructure will satisfy customer needs while being compatible and complementary to growing rocket portals. Each will have strengths to support various customers; however, the movement of cargo for complex and massive undertakings is a natural strength of Galactic Harbours. Complementary transportation portals and infrastructures can ensure success for different missions and destinations desired by future movements off-planet.

Leveraging information about demand pull for these reference missions and comparing carrying capacity of early, and then mature, Space Elevator transportation infrastructure results in Table 7.1: Fulfillment of Reference Missions. This puts the whole picture into focus. The demands are huge for these critical reference missions and their destinations. FOC Galactic Harbours are needed as soon as possible to support humanity's dreams. This vertical logistics chain can enable colonies on Mars, Lunar Villages and Space Solar Power.

Table 7.1, Galactic Harbour Fulfillment of Reference Missions

Reference Mission	Metric tonnes to Destination	Galactic Harbour IOC Fulfillment Time (yrs)	Galactic Harbour FOC Fulfillment Time (yrs)
Space Solar Power	5,000,000	150	29
Mars Colony	1,000,000	33	6
Moon Village	500,000 estimated	17	3

7.7 Unique Strengths are Required:

In addition, the unique characteristics of release from Apex Anchors leads to the conclusion that interplanetary flight and movement off-planet requires a complementary infrastructure with rockets and Space Elevators sharing missions with each system leveraging their strengths. Rockets are good for near-Earth missions and some manned missions because they are fast; but, for interplanetary missions, Space Elevators win hands down. Of course, all missions need rocket engines to slow down and rendezvous with destinations. The new concept of Galactic Harbour Architectures has unique characteristics that will "enable" interplanetary missions as complementary infrastructure to rockets. These include:

- **Routine Massive Lifts:**   During early operations, each Space Elevator Climber will carry 14 metric tonnes of payload to GEO and beyond with departures every day, or 84 metric tonnes per day (14 x 2 SE x 3 GH) around the globe. This will happen 365 days a year, or 30,660 metric tonnes per year to GEO and beyond. As the maturity is reached in massive liftoff Space Elevators, the number moves up to just less that 200,000 metric tonnes per year to GEO and beyond (see chapter 5 for development of these numbers).

- **Routine Daily Lifts:**   As the Space Elevator is designed to lift cargo daily, releases towards interplanetary missions will be standard and routine.

- **Fast Transits to Mars Available:**   With the daily release of payloads towards Mars (and other interplanetary destinations) release from the Apex Anchor imparts tremendous velocity with very little drag from Earth's gravity. As a result, a periodic fast transit to Mars lowers the minimum time to 61 days.

7.8 Change in Off-Plane Approach:

This ISEC study report results show that there must be major changes in the approach for humanity's migration off-planet. Some of these changes should include:

- Change of vision for interplanetary movement when delivery of mass is inexpensive, timely, environmentally friendly, daily, and supportive. It turns out that the revolution in transportation capabilities of Galactic Harbours opens up immense possibilities and ensures that humanity can "bring with them" the essential elements for survival and aggressive growth. This new vision of Galactic Harbour architectures will change the thinking for off-planet migration - We CAN bring it with us!
- Movement off-planet will require complementary capabilities, such as rocket portals and Galactic Harbour infrastructures, each with their own strengths and

short-falls. Moving people through the radiation belts rapidly is a strength of rockets while massive movement in a timely, routine, inexpensive and Earth friendly manner are the strengths of Space Elevators.

- This discussion of mission needs when analyzing mass to location will start the analysis of "how much carrying capability" is required by each supportive infrastructure; when, to where, and their priorities. In the past, the rocket approach valued light-weight, compact, designs of support equipment while the Space Elevator permanent infrastructure will enable mass to be moved to desired destinations easily. The driving function for infrastructure design becomes a description of the customers' needs, not light weight designs.

Some conclusions from these analyses are stimulating!

- The Space Elevator is Closer than you think.
- Galactic Harbours will enable robust missions to Moon and Mars!
- Only Space Elevators can deliver the requirements of logistics equipment and supplies to the Moon and Mars
- Colonization on Mars cannot happen without the logistics support of Space Elevators!
- Launch Windows to Mars every 26 months - be Damned! [a favorite]

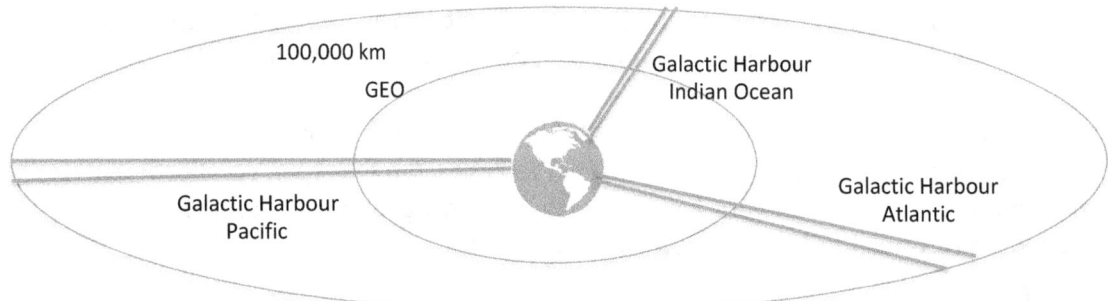

Figure 7.3, Three Galactic Harbours - A Vision At the Horizon

Chapter 8 - Recommendations

8.1   Recommendations:

Near term activities for Space Elevator development will focus on transition from Engineering Development Phase One (Technology Feasibility) to Phase Two (Engineering Validation) as the activities shift from ISEC, an advisory association, to industry where engineering testing can be conducted. Expansion on this is available in Appendix C. In parallel with the transition of engineering activities, there would be four major steps to be initiated or continued as the program moves forward:

8.1.1   Near Term Initiative ONE: Create a Space Elevator Institute: Beyond the engineering refinements occurring across the industry, there is one major step that must be initiated in the near term to significantly move the Space Elevator / Galactic Harbour concepts forward. A Space Elevator Institute should be established to look at the newly emerging mosaic of space and how Space Elevators enable the future. The charter of a Space Elevator Institute would encompass the issues for discussions, research, analysis, and policy; and formulate recommendations for execution. Establishing an Institute that will be the leader in all things space elevator. The codification of this engineering transportation infrastructure will solidify the segment to segment relationships and support the satisfaction of system level requirements in preparation of Space Elevator / Galactic Harbour design and developmental activities.

A second thrust inside this initiative needs to focus on investigations into such areas as funding approaches, relationships with supporting governments around the world, discovering its rightful place in interplanetary support activities, and determining best approaches to develop Enterprise Infrastructures across Galactic Harbours. The ability to assess current and near term activities with a historical view will make this concept of a space mosaic even more enticing to future generations.

In addition to starting and conducting research across the disciplines of Space Elevators, the Institute should also have a short-term goal of flying a Pathfinder Space Elevator (1,000 km long tether at 2,500 km altitude orbit). This Pathfinder will illustrate the maturity of various technologies within all of its major segments. (see Initiative Four) Once the Pathfinder has succeeded, major funding for a full up Space Elevator should be initiated and lead to the full sequence of construction of the first Initial Operational Capability transportation infrastructure.

8.1.2   Near Term Initiative Two: Support and encourage Further Research at Arizona State University: Much of the student work (2019 & 2020 classes) developed the code for understanding this new capability of release from a Galactic Harbour's Apex Anchor. The focus was on Mars with some work looking at going to Jupiter's moon, Io. Recent work, in the spring of 2020, with a new set of students from Professor Peet's orbital class, took on new tasking. Covid-19 had a major impact at all universities, and ASU was no exception. Therefore, these new topics are not be covered in this report. Some future ideas that could be pursued in future classes are:

- Apex Anchor to Mars (with a receiving Space Elevator rendezvous)
- Apex Anchor to the Moon (with a receiving Space Elevator rendezvous)
- Trips to the Inner Solar System - Venus and Mercury as well as smaller solar orbits.
- Science missions to the outer planets
- Science missions leaving the solar system (with inherent escape velocity from a 163,000 km altitude Apex Anchor)
- Alternative approaches to gaining extra velocity - trebuchets and/or slide release

Additional Destinations: The flexibility to support various missions to diverse destinations is a tremendous strength of Galactic Harbours. Further study looking at the various velocities at release from an Apex Anchor will show how Galactic Harbours enable flights to all the destinations of choice within our solar system (and even beyond with the proper additional propulsion or gravity assists).

Table 8.1: Height of Apex Anchor Alternatives

Length (kms)	Author	Speed (Km/sec)
96,000	Obayashi (2013)	7.466
100,000	Edwards (2002), IAA (2013 & 2019)	7.757
120,000	Knapman (2019)	9.216
144,000	Pearson (1975)	10.966
150,000	ASU alternative (2018)	11.403
163,000	Swan (2020) Solar System Escape Velocity	12.352

8.1.3 Near Term Initiative Three: Support and encourage testing of newly discovered 2D strong material: This activity should have emphasis from all of the Space Elevator researchers and supporters. The development of single crystal graphene (or equivalent 2D material) is critical to the development of space elevator tether; as such, massive support must be encouraged. The next chart shows a research approach resulting in a capability to have continuous manufacturing (fast enough) to provide long tethers with great strength.

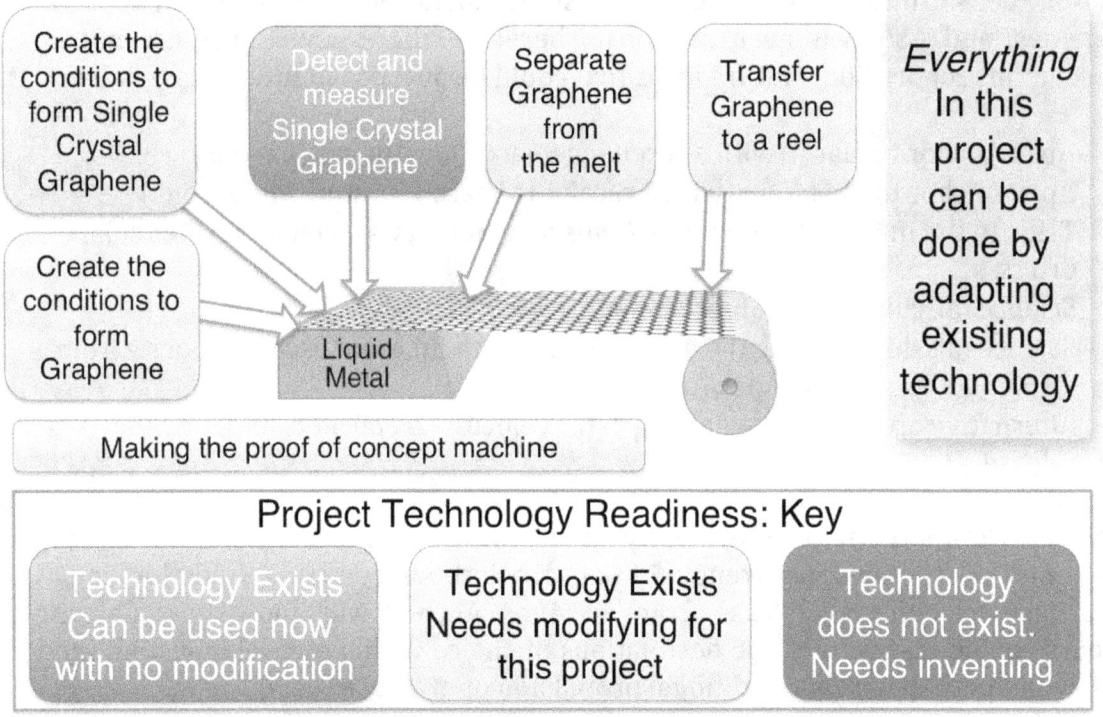

Figure 8.1, Continuous Single Crystal Graphene Manufacturing:
Project Technology Readiness (Slide from Adrian Nixon's presentation)

8.1.4   Near Term Initiative Four: Testing in space: After all, this is a space transportation infrastructure:

Pathfinder Demonstration: After all the culminating demonstrations across the five segments, the validation for the customer will be a single full up system of systems integration test that shows the interplay across the segments as well as within each segment. There is a concept surfacing within the Space Elevator community that sees a full-up test in space once the culminating demonstrations have been accomplished with the individual segments. This would be designed to show to customers that the concept works in the environment of a Space Elevator, just at a smaller scale prior to commitment of the final investments. The concept is to have a long [estimate 1,000 kms] tether [of the same material but less strength than required for the Space Elevator] being placed at 2,500 km altitude circular orbit. The initial spacecraft would resemble the future deployment satellite with both a smart satellite representing the Earth Port and a smart satellite representing the Apex Anchor, connected by a 1,000 km tether. The system of system integration test would validate the following actions to ensure design maturity and concept viability.

- Placement in Orbit (into LEO then transfer to 2,500 km circular)
- Stabilization in circular orbit [2,500 km altitude]
- Deployment of a tether under control [Apex Anchor deploys Earth Port under conditions similar to reality with HQ/POC controlling]
- Full tether deployment [1,000 km length]
- Apex Anchor Controls dynamics with thrusters and reel-in reel-out capabilities
- Earth Port controls dynamics with thrusters and reel-in reel-out capabilities
- Tether Climbers are deployed and climb up and down with mechanisms representing the full up design of space elevators. In addition, they would monitor the tether for wear due to climbers, radiation, particles and debris.
- All actions controlled and monitored from the HQ/POC primary center.
- Release of tether climbers as last test [of course release to deorbit for debris cleanup]
- Move system to deorbit path.

Once the execution of this system of systems integration test is completed, most of all the serious questions of space elevator operations will have been addressed and answered showing a positive situation ready for commercial funding to finish the first Space Elevator.

8.2   Galactic Harbour 2020 Vision:

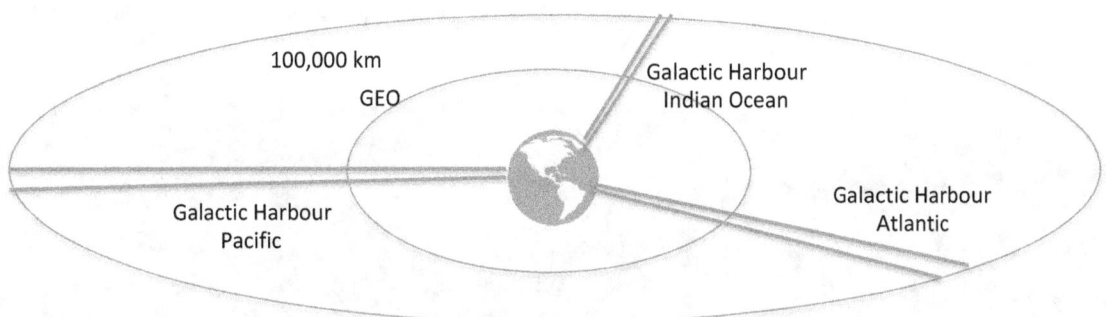

The Space Elevator story is still being written.
The Apex is where the Galactic Harbour meets the Shoreline
of Outer Space and Where the "Transportation Story of the
21st Century" meets the "Final Frontier."

References[48]

- Artsutanov, Y. V Kosmos na Electrovoze, Komsomolskaya Pravda, July 31, 1960.
- Artsutanov, Y. To the Cosmos by Electric Train, SF magazine, Vol. 2, No. 2, pp121-123, 1961
- Edwards, B. & E. Westling, The Space Elevator, BC Edwards, Houston, 2002.
- Fitzgerald, M, R. Penny, P. Swan, C. Swan, Space Elevator Architectures and Roadmaps, ISEC Study Report, lulu.com, 2015
- Fitzgerald, Michael, Vern Hall, Cathy Swan, Peter Swan, Design Considerations for Space Elevator Apex Anchor and GEO Node, ISEC Study Report, lulu.com, 2017.
- Fitzgerald, Michael, Technical Maturity and Development Readiness of the Galactic Harbour, IAC-19, paper and presentation, Washington D.C., Oct 2019.
- Fitzgerald, Michael, Architectural Note 1 thru 31: various topics www.isec.org Architect Page
- Hall, Vern, R. Penny, P. Glaskowsky, S. Schaeffer, Design Considerations for Space Elevator Earth Port, ISEC Study Report, www.lulu.com, 2016.
- Ishikawa, Yoji, The Space Elevator Construction Concept, Obayashi Corporation, 2013, IAC-13-D4.3.6.
- Kemble, Stephe, Pg 9, Interplanetary Mission Analysis and Design, pg 9, Springer, 2006
- Knapman, John, P. Glaskowsky, D. Gleeson, V. Hall, D. Wright, M. Fitzgerald, P. Swan, Design Considerations for the Multi-Stage Space Elevator, ISEC Study Report, lulu.com, 2018.
- Mankins, John, The Case for Space Solar Power, Virginia Edition Publishing Co. Dec 2013
- McFall-Johnsen, Dave Mosher, "Starship rockets every day and creating 'a lot of jobs' on the red planet," Jan 17, 2020, Business Insider.
- Musk, Elon, "Making Humans a multi-Planetary Species," New Space, Vol 5,No2.
- Pearson, Jerome , The orbital tower: a spacecraft launcher using the Earth's rotational energy, Acta Astronautica. Vol. 2. pp. 785-799. Pergamon Press 1975. Printed in the U.S.A.
- Penny, Robert. Swan, Peter, & Cathy Swan, "Space Elevator Concept of Operations," ISEC Position Paper #2012-1, International Space Elevator Consortium, Fall, 2013.
- Penny, R., P. Swan, C. Swan, J. Knapman, P. Glaskowsky, Design Considerations for Space Elevator Tether Climbers, ISEC Study Report, www.lulu.com, 2014

[48] Complete list of references and citations about space elevators on website: www.isec.org

- Raitt, David, Space Elevators: A History, ISEC Report 2017.
- Swan, P., Raitt, Swan, Penny, Knapman. International Academy of Astronautics Study Report, Space Elevators: An Assessment of the Technological Feasibility and the Way Forward, Virginia Edition Publishing Company, Science Deck (2013) ISBN-13: 978-2917761311
- Swan, P., David Raitt, Space Elevator – 15 Year Update, Journal of British Interplanetary Society, Vol 69, No 06/07, Dec 2016.
- Swan, P., David Raitt, John Knapman, Akira Tsuchida, Michael Fitzgerald, Yoji Ishikawa, Road to the Space Elevator Era, Virginia Edition Publishing Company, Science Deck (2019) ISBN-19: 978-0-9913370-3-3
- Swan, P., Michael Fitzgerald, Today's Space Elevator, ISEC Study Report, lulu.com, 2019.
- Swan, Peter, Robert "Skip" Penny, and Cathy Swan, Space Elevator Survivability – Space Debris Mitigation, Lulu.com, 2011.
- Torla, James and Matthew Peet, Space Elevator Support for Interplanetary Flight. Presented at NSS International Space Development Conference, Washington, D.C. June 7-9 June, 2019.
- Torla, James and Matthew Peet, Optimization of Low Fuel and Time-Critical Interplanetary Transfers using Space Elevator Apex Anchor Release: Mars, Jupiter and Saturn, IAC-19, paper and presentation, Washington D.C., Oct 2019.
- Tsiolkovski, K. E., Grezi o zemle I nebe, izd-vo ANSSSR, 35, 1959.
- Wang, X., B. Shen, Y. Xu, K. Tong, L. Shen, Y. Lu, "Study on a small-scale and high-performance space elevator," IAC-18 D4.IP.ll, 2018.
- Wright, Dennis, S. Avery, J. Knapman, M. Lades, P. Roubekas, P. Swan; Design Considerations for a Software Space Elevator Simulator, ISEC Study Report, lulu.com, 2017
- Zhaohui, G., K Tong, F. Zhang, Y., Cai, "Mission Analysis of Human Mars Exploration based on Space Elevator," Proceedings of 2016 IEEE Chinese GN&C Conference, 2016.

Appendices

Appendix A: Defeating the Rocket Equation

The Bottom Line: It is important to remember, Space Elevators are compatible and complementary to rocket architectures. The future needs both communities to work together. However, the first step is to help the rocket community understand the strengths of space elevators - "we can beat the rocket equation."

Problem: The Earth's gravity well has enabled life to develop. The atmosphere is dense, the movement of tectonic plates have enabled diversity and evolution, and the Moon supports us each day with a bright light and ocean tides. However, to escape, from the huge home base gravity well, has limited our migration off Earth. There is a lot to the gained from the defeat of adversaries. In this case, gravity. Our escape to the Moon for Apollo and the ability to pursue science in all parts of our solar system has demanded our best efforts and exceptional engineering feats. One over radius squared, from a big planet, is a difficult problem that needs to be defeated.

Examples: To escape is the first step of moving off planet (Low Earth orbit - achieved in 1957). It required the best of engineering feats by the Soviet Union. All space efforts since then have required huge masses of fuel and structure to leave the planet and gain orbital positions. This is usually explained in the terminology of gaining enough velocity to stay in orbit. To gain LEO, the accepted value is 9.3 km/sec velocity gained by burning fuel. Here lies the problem: We must burn fuel and send it out as exhaust to move the mass of the vehicle forward. Over the years, the consumption of 96% of the mass that starts on a launch pad is thrown away as the "cost of doing it this way." This included all the fuel needed to burn and push the rocket, the structures to hold the fuel, the rocket nozzles, and all the other structure needed to hold the payload safely in its grasp. We can all discuss the numbers, but a reasonable assumption is 4% of the mass on the pad gets to Low Earth Orbit. Another example was delivery of the Apollo lunar lander to the surface of the Moon was estimated at half of one percent of the launch mass (0.5%) reached the lunar surface.

Note: The reusability of stages and the cost effective approach that is being refined and exercised by all the rocket companies (lead by Blue Origin and SpaceX) are very good at being more efficient and even more cost effective - lowering the costs to orbit into a range of costs never expected. However, when one looks at the Tsiolkovsky rocket equation, one does not see any factor with reference to cost or reusability. As such, the mass to orbit is still not any more than the old numbers of 4% of pad launch mass. Goddard and Von Braun recognized this monumental problem and found ways to "work through it." An estimate of the SpaceX StarShip rocket capability is 100,000 kg to LEO with pad launch mass of 5,000,000 kg (estimates on wiki - 22 June 2020). This leads to only 2.0 % of launch pass mass to LEO. The system is very efficient and cost effective; however, it has not beaten the rocket equation and "big gravity."

The Rocket Equation[49]:

"The Tsiolkovsky rocket equation, classical rocket equation, or ideal rocket equation is a mathematical equation that describes the motion of vehicles that follow the basic principle of a rocket: a device that can apply acceleration to itself using thrust by expelling part of its mass with high velocity can thereby move due to the conservation of momentum. The equation relates the delta-v (the maximum change of velocity of the rocket if no other external forces act) to the effective exhaust velocity and the initial and final mass of a rocket, or other reaction engine. For any such maneuver (or journey involving a sequence of such maneuvers):"

$$\Delta v = v_e \ln \frac{m_0}{m_f} = I_{sp} g_0 \ln \frac{m_0}{m_f}$$

where:

Δv is delta-v – the maximum change of velocity of the vehicle (with no external forces acting).

m_0 is the initial total mass, including propellant, also known as wet mass.

m_f is the final total mass without propellant, also known as dry mass.

$v_e = I_{sp} g_0$ is the effective exhaust velocity, where:

I_{sp} is the specific impulse in dimension of time.

g_0 is standard gravity.

$\ln$ is the natural logarithm function.

The words of consequence are: "a device that can apply acceleration to itself using thrust by expelling part of its mass with high velocity can thereby move due to the conservation of momentum."   The Tsiolkovsky rocket equation still responds to that critical factor called gravity. The Earth's gravity numbers have a consistent impact on effectiveness at liftoff and flight - DRACONIAN!

---

[49] Wikipedia 22 June 2020

Space Elevator Strengths: For the GEO Region and beyond (including all solar system destinations) the Space Elevator "Beats the Rocket Equation." How is this done? Simple - it raises the cargo for each destination up to an altitude using electricity - not consuming rocket fuel and structure. As a result, the payload of the tether climber gains energy from the process. Using the Apex Anchor location as an indicator of the process, the payload has gained 100,000 km of potential energy and results in a horizontal velocity of 7.76 km/sec at release. This resulting energy gain has "enabled" the payload to go to Mars any day of the year (no waiting for 26 months for a launch window) and as rapidly as 61 days to Mars (over 200 releases across the planet's periodic dance of less than 200 days with many at 75-90 days to Mars). This is all achieved because the space elevator lifts the payload out of the gravity well and releases it when gravity is very low.

This image shows the various release velocities at the lengths of space elevator tether. If one were to release at the Geosynchronous altitude, the payload would go into a geosynchronous orbit. As the height of release goes up, the velocity at release increases. The currently conceived length of a space elevator tether is 100,000 km; and, as such, provides enough velocity to reach the Moon in 10 hours or Mars in as little as 61 days. If one were to go to a 163,000 km altitude on a space elevator and release from the Apex Anchor, the payload could escape the Solar System without additional thrust (of course the mission would probably use gravity assist to gain velocity and rockets to correct the trajectory as it traversed open space).

12.35 km/sec
163,000 km

11.4 km/sec
150,000 km

7.76 km/sec
100,000 km

3.078 km/sec

Geosynchronous
Altitude

Space Elevator
Launch Geometries[50]

Conclusion #1: How can this be possible? Simple - a Space Elevator infrastructure need only defeat gravity and the traditional rocket equation once. Massive payloads to the Apex Anchor - raised by electricity - to be released at 7.76 km/sec towards destinations; daily, routinely, safely, and robustly all while being environmentally friendly.

---

[50] Torla, James and Matthew Peet, "OPTIMIZATION OF LOW FUEL AND TIME-CRITICAL INTERPLANETARY TRANSFERS USING SPACE ELEVATOR APEX ANCHOR RELEASE: MARS, JUPITER AND SATURN," International Astronautics Congress (IAC-18-D4.3.4), Washington D.C., 2019.

Conclusion #2:   ISEC has shown that there must be major changes in the approach for humanity's migration off-planet. Some of these changes include:

- Change of vision for interplanetary movement when delivery of mass is inexpensive, timely, environmentally friendly, daily, and supportive. It turns out the revelations in transportation capabilities of Space Elevators open up immense possibilities and ensures that humanity can "bring with them" the essential elements for survival and aggressive growth. This new vision of Space Elevator architectures will change the thinking for off-planet migration - We CAN bring it with us!

- Movement off-planet will require complementary capabilities -both rocket portals and Space Elevator infrastructures - each with their own strengths and short-falls. Inserting payloads into Low Earth Orbits and moving people through the radiation belts rapidly are strengths of rockets while massive movement in a timely, routine, inexpensive and Earth friendly manner are the strengths of Space Elevators.

- A discussion of various destination mission needs when analyzing mass to location, will start the analysis of "how much carrying capability" is required by each supportive infrastructure: when, to where, and their priorities. In the past, the rocket approach valued light-weight and compact designs of support equipment while the Space Elevator permanent infrastructure will enable mass to be moved to desired destinations easily. The driving function for infrastructure design becomes a description of the customers' needs, not light weight designs.

- An interesting insight in parallel with this analysis says that planetary scientists can be offered as much mass as they require for any of their missions. There will be zero restrictions for scientific instruments going to any place in the solar system - including the survival from the shake, rattle, and roll of rocket launches.. If you can not include it in one 14 metric ton payload capable tether climber, you can assemble parts at the Apex Anchor and release them once a day towards any destination.

Appendix B: 2020 Vision

"The Space Elevator enabled network of Galactic Harbours"
Vernon Hall & Michael Fitzgerald[51]

2020 Vision. It sounds like a diagnosis from your optometrist; but it isn't. It is what the Space Elevator, the Galactic Harbour, is becoming. Our 2020 Vision is a portrayal of the fulfilled transportation story of the 21st Century. It is the extension of our experience; the manifestation of humankind's initial expansion into the rest of the Universe. This is an unabashed explanation of what we see with 2020 foresight.

An architectural engineer is a system engineer with vision like Michael has. You have to completely rebuild a modern harbor to understand the "mega" part of megaproject like Vern did. These co-founders see a network of space elevators as a visionary megaproject; one that is the culmination of our vision and our life's work.

In the broad field of Transportation Systems, the 19th and 20th Centuries saw *evolutionary* advances in the movement of cargo and people across vast distances. In the United States, the opening of the *Transcontinental Railroad* in the 1860s brought expansion and opportunity while making such systems as covered wagons, pony express and even "around the Horn" sailings obsolete. In the early 20th Century, gasoline powered automobiles and trucks led to development of roads and highways that expanded into the *Interstate Highway System* in the 1950's. In this same era, visionaries developed the concept of *Intermodalism* based on standardized shipping containers; soon becoming the worldwide *Containerization* system. That system remains the primary mover of large cargo volumes across the face of the Earth. In the 1970s, FedEx developed its system of specialized *Air Cargo* planes that revolutionized the movement of consumer goods.

By the middle of the 21st Century *Galactic Harbours* will provide a nexus for connecting these two dimensional transportation systems of rail, highway, shipping, and air to the third dimension; and provide a safe, reliable and cost effective means to move material and people into orbit and beyond.

Earth's Galactic Harbours

We see a network of diverse elevators, spread around the earth; but they differ by more than location. Their respective engineering differs by their purpose. Cargo

---

[51] *Michael Fitzgerald and Vern Hall are two Americans living in Southern California, USA. They met over 30 years ago and began collaborating on the Space Elevator about 5 years ago. Vern's career centered on remaking the Port of Los Angeles into the major trade and transshipment center in the United States. Michael has spent nearly 50 years in the engineering development of major space projects. Their partnership and their work developing the Galactic Harbour, is the basis for the Vision they present here.*

transshipment and vertical cargo transport are common to each, but the Harbours are optimized for their respective specific purposes.

- GEO Orbit factory support
    - Space Based Solar Power
    - Satellite servicing – Refuel, repair, upgrade
    - Assembly of large spacecraft and space facilities.
    - Other on orbit enterprises
- Interplanetary travel support
    - Flight Systems final assembly
    - Departure Mission readiness and checkout
    - Departure and arrival operations
    - Fueling special thrusters for long distance operations
    - Massive cargo and material handling operations
- Research support
    - Sensing Systems
    - Flight operations within the Elevator regions
    - Computers
    - Testing and test data collection facilities
- Tourism support
    - Human rated activity
    - Elevator region flight operations
    - Safety and Comfort
- Government support
    - Space and Region debris mitigation
    - Space Traffic Management and Control
    - Law enforcement
    - First Responders
    - Safety assurance operations
    - Sovereign issues and relations

Some of these elevators are stand alone, and some are hybrid elevators sharing the operational regions with one another. To some extent the "Government elevator" is embedded in all elevators. Travel from one elevator to another is on enhanced pathways along and across the geosynchronous belt. Trade between the elevators is robust; especially for delivery of key repair items; FedEx and UPS aloft. Some of the elevators are human rated and some are purely robotic. Some are small and speedy, and some have immense throughput.

The Solar System's Elevators

The purposes of Earth's Space Elevators are to 1) deliver cargo to the Enterprises assembling along the geosynchronous belt near the Space Elevator's GEO Regions;

and 2) support interplanetary flights from the APEX Regions; to the Moon, to Mars, and elsewhere.

Since the interplanetary flights from the Apex Anchor will use the latent ΔV, (derived from Earth's rotation - transformed into radial speed at departure from the Apex); Space Elevators are established around the Solar System as part of the Galactic Harbour transportation network. Elevators operate near the Moon, on Mars, on key asteroids within the asteroids belt and elsewhere. We see immense cargo craft moving from Elevator to Elevator bringing supplies and equipment; and returning with raw materials for processing in one of the several GEO regions and later to Earth. This is the third dimension of trade, commerce, transportation, and humankind.

The Galactic Harbour Network –
Earth's lifeline to the future

The magnitude of this portrayal is humbling. It will be accomplished in the time to come. It is a well-marked destination; marked by the needs of humanity. It was seen before; yes, seen by those who preceded us - and it will be built by those who follow.

Appendix C - Space Elevators Ready to Proceed: Significant Q&A

C.1 Introduction: This study report is being published at a time when the space elevator is "breaking out" of its old mode of technical progress and moving into an aggressive approach to be included in the serious discussions.   The following major questions are answered in this appendix with references to further support. A key to this discussion is;

The Space Elevator is Closer than one Thinks!

The questions being asked around the community that have already been answered by International Space Elevator Consortium, Japanese Space Elevator Association, and the International Academy of Astronautics are:

- What is the current schedule as seen by IAA and ISEC?
- Where is the development of Space Elevators inside Engineering Stages?
- What is the material for the tether and where is its development?
- How will the Multi-Stage Space Elevator concept enable early start?
- Do we have a "gold standard" for simulations?
- Assured Space Debris Survivability?

C.2   What is the current schedule as seen by IAA and ISEC?

ISEC Study Reports and the International Academy of Astronautics Study Reports have stated that the Preliminary Technology Readiness Assessment has been achieved and the Space Elevator should be developed towards operations to enable significant movement off-planet. The initial operations should be somewhere around 2035. The latest ISEC Study Report, *Today's Space Elevator*, has a suggested schedule that looks out to Initial Operational Capability and then beyond to Full Operational Capability.

Event Occurring between two dates	Early	Estimated
Material for Tether shows Characteristics	2019	2021
Material developed for Space Elevator Tether	2023	2029
Major Segments Validation Testing	2024	2030
Integrated Orbital Testing (Low Earth Orbit)	2031	2032
Launch of Deployment Satellite	2032	2034
Deployment of Space Elevator	2033	2036
Buildup of Space Elevator to Initial Operations Capability	2037	2040
Initial Operations	2037	2040
Galactic Harbour Operational	2037	2040
Second Galactic Harbour Operations	2039	2042
Full Operations Capability (with People)	2047	2057

Table 2, Space Elevator Proposed Schedule[52]

C.3   Where is the development of Space Elevators inside Engineering Stages?

This recognition of a "Sea State Change" has driven the community to illustrate the engineering stages for each segment of the Galactic Harbour architecture. This naturally leads to the need for others to recognize these approaches and successes.

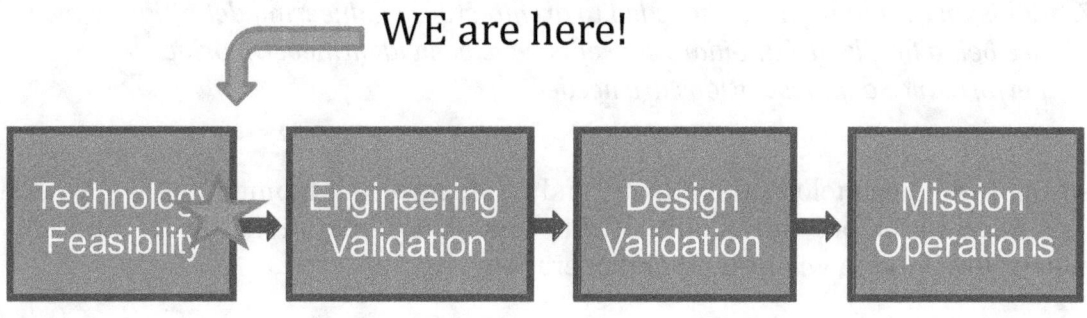

Figure 10, Space Elevator Level of Maturity

[52] Swan, today's

This requirement to "show the numbers" was met over the last ten years with constant analyses and specific study topics to expand Space Elevator knowledge. The following few paragraphs show how the Space Elevator engineering status has moved past the Preliminary Technological Assessment towards development and operations.

In the last six years, ISEC's technology maturation approach has melded with a better definition of Space Elevator engineering solutions. The 2014 publication of ISEC's "Architecture and Roadmap" report removed the shroud of mystery and myth from the elevator's scope and complexity. The Space Elevator was no longer a mystery. "Design Consideration" documents published between 2013 and 2017 delineated an engineering approach for Tether Climber, Earth Port, GEO Region, and Apex Anchor. An architectural simulation tool was selected. The last technology hurdle - a strong material for the tether – is being overcome. The figure below shows where Space Elevator engineering is within the developmental process.

To illustrate the closeness to exiting Technology Feasibilty, ISEC has laid out the summary of each major segment, showing that the Space Elevator program is ready for Engineering Validation. The ISEC position shows:

1. *The Galactic Harbour Earth Port is ready for an engineering validation program*
2. *An engineering validation program for the Space Elevator Headquarters / Primary Operations Center is ready to begin.*
3. *Tether Climbers development should start with engineering model assemblies, followed by an engineering validation program.*
4. *GEO Node engineering discussions and demonstrations can be accomplished with key members of industry and collaboration / outreach with certain government offices.*
5. *Apex Anchor is in the middle of engineering discussions and various simulations. Near term collaboration with engineering organizations and academia should be started.*
6. *Tether material is now "real." A prime material candidate has been identified. Production demonstrations are needed.*
7. *Collision avoidance needs have led to architecture engineering definitions which are being finalized. Candidate concepts have been identified. On orbit performance demonstrations are needed.*

Based upon this technological maturity, and its engineering momentum, we expect that significantly before the middle of this century an operational Space Elevator Transportation System will be built and operating.

The Technology Momentum of the Galactic Harbour is real; and, it underwrites the interplanetary vision of transportation, enterprise, and exploration

This discussion addresses a milestone for all members of ISEC. It documents, in summary form, how we have entered the Space Elevator Era. The discussion explains how we moved from a roadmap study in 2014 to today -- ready to declare "Tech Ready." All readers should at least examine the graphics and understand them. Industry must now get involved and ISEC should help them. The majority of this discussion has come from ISEC's Architectural Note #24, The Path to Tech Readiness[53].

The technology momentum of the Galactic Harbour is real; and it underwrites the interplanetary vision of transportation, enterprise, and exploration. In the last year, the International Space Elevator Consortium asserted that the basic technologies needed are available; and, that each segment of the Space Elevator Transportation System is ready for engineering validation. The ISEC position has developed as follows:

C.3.1   The Space Elevator is nearing the end of Technology Development Phase

During the 2014 road mapping effort, it seemed evident that within the envisioned Space Elevator Architecture, new entities and technologies would be required, engineering approaches would need refinement, and new materials would need to be found. A technology development approach was built based upon a development approach of "Show Me." They were essentially a set of well-constructed demonstrations, inspections, tests, simulations, experiments and analyses - best conducted by industry (as industry will be building the elevator). ISEC also noted that each segment of the elevator has its own challenges and will likely need to resolve those challenges in segment unique manner. The technology and engineering issues facing something at the equator and in the middle of the Pacific Ocean are not directly relatable to something at the APEX Anchor at 100,000 kilometers above the equator in the middle of "Outer Space." As much as the issues are dissimilar; they are the same. They must be defined and their solutions found. After that it must be determined if something can be built from them. The work retains that theme. ISEC's technology development follows a tried and true sequence. The approach extends the thinking of industry / commercial Technology Plans.

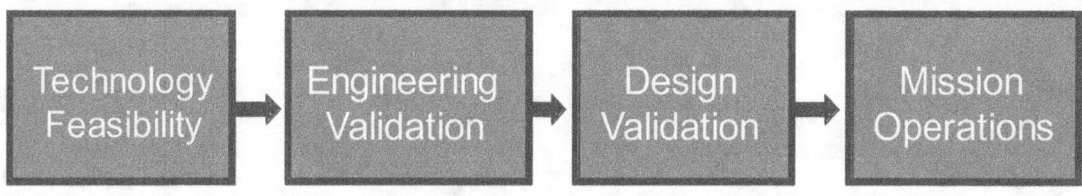

Figure 6, Engineering Developmental Phases

---

[53] Fitzgerald, M. aRch note 24

The progress within the plan continues to be based upon an iterative approach to risk mitigation. Recurring Technology Readiness Assessments culminating in operations demonstrations & prototypes such as success at Initial Operational Capability are required54.

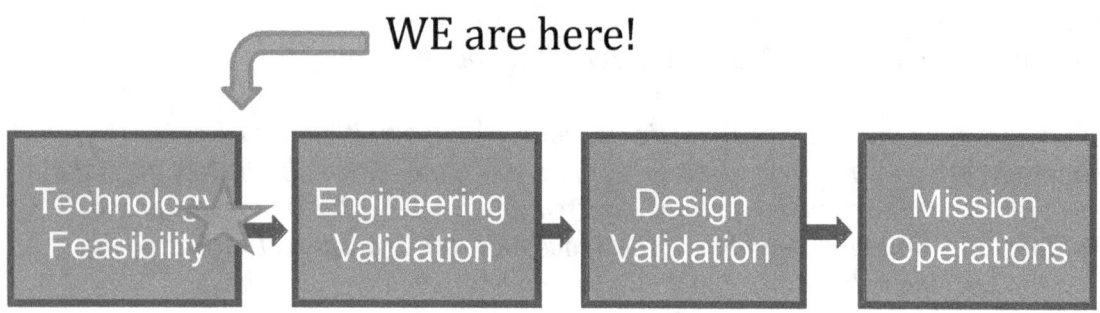

Figure 7, We are Here, Between Phases

The ISEC team has been assessing the technology feasibility situation since 2008. In recent times, the team has begun an open dialog with those members of industry, academia, and others; who could be the deliverers of ISEC solutions. Industry (especially) will show how the needed technologies are being matured and when they could be dependably available.   These readiness assessments are the Phase One exit criteria:

- *Document technology readiness state. Determine if the technologies are State of Art (SOA) or State of the Industry (SOI) or State of the Market (SOM)*
- *Establish readiness level rationale for all portions of the Program. Given that the technology availability has been demonstrated the level of readiness can be established for each program segment*
- *Set Success Criteria regarding Engineering Validation – the second phase. Prudent acquisition planning calls for an early design reviews. "Show me" means a lot at this point.*

---

54 Fitzgerald, Michael, Space Elevator Pathway to Technology Maturity … and Beyond, From Fountains to Tech Ready. presented at 2019 International Space Elevator Conference, Seattle, 16-18 Aug 2019.

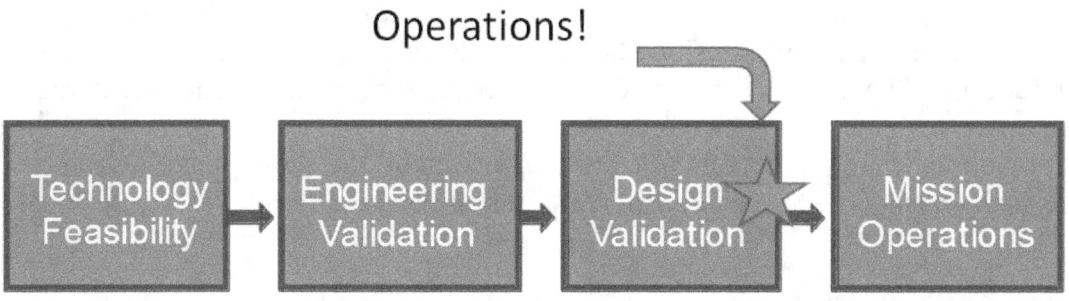

Figure 8, Operations is Downstream, the Fourth Phase

C.3.2   Phase two: Validate engineering approaches

Phase two will begin soon after phase one completion. Industry involvement is an imperative. Phase two activities are driven by six major activities:
- *Examine Industry's production foundation*
- *Determine if the segments can be built*
- *Assess schedule & technical risk*
- *Delineate design criteria*
- *Set criteria and standards to enter the Design Validation Phase*
- *Baseline operations performance:*

C.3.3   Engineering Development Status as of Summer 2020:

Because of these achievements, there are many questions that have been answered over the last ten years with respect to the readiness of Space Elevators. The position of ISEC is that the:

> *Space elevator development has gone beyond a preliminary technology readiness assessment and is ready to enter initial engineering validation testing -- leading to the establishment of needed capabilities.*55

C.4   What is the material for the tether and is it ready?

ISEC believes a material is in the laboratory that approaches the 150 GPa tether requirement. Although the IAA/ISEC approach needs much less than the old requirement, the new material [Single Crystal Graphene] could be manufactured in

---

55 Swan, Peter, Michael Fitzgerald, "Today's Space Elevator," ISEC Study Report, lulu.com, 2019.

long lengths as "single crystals.[56]" (130 GPa, with a density of 2.2g/cc). The minimum requirement is 84 GPa at this density. The estimate is for long tether material strong enough for Space Elevators is the late 2030's. A recent letter from Nixene Ltd. stated:

"Joint planning between ISEC and Nixene Ltd anticipates the development testing and deployment of the Space Elevator tether within the next decade or two...."[57]

The following is from "Today's Space Elevator" (expanded Adrian Nixon charts)

Conclusions: The tether material is the pacing item for the development of a Space Elevator. Currently, there are three materials that could grow into the needed strong-enough and long-enough material for a Space Elevator: carbon nanotubes, boron nitride nanotubes, and continuous growth graphene. The community waits for those materials to mature to the level that can be implemented into a Space Elevator tether 100,000km long and strong enough to support its own weight plus multiple tether climbers against the pull of gravity. At 100,000 kilometers long, a Space Elevator tether is a major engineering challenge. Recent investigations explored the possibility for making single crystal graphene by a continuous process using liquid metal. Making this a viable practical manufacturing process will be a significant effort over a period of years and probably many millions of dollars. However, such a process would create graphene products for many multi-billion dollar markets on the way to making the tether material. For this reason there is a credible return on investment case for manufacturing the material. This means it is highly possible that continuous single crystal graphene will be manufactured in the coming years and this material should be considered in any forward thinking about Space Elevator tethers.

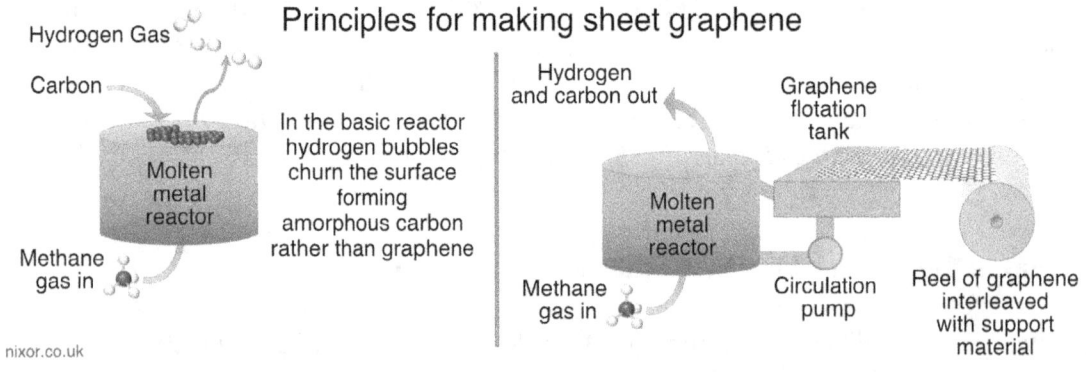

Figure xx, Overcoming CVD production problems: A continuous process

---

[56] Nixon, Adrian, Update on Graphene as a Tether Material. 2019 International Space Elevator Conference, Seattle, 16-18 Aug 2019.
[57] Adrian Nexion letter

(Slide from Adrian Nixon's presentation webinar June 2019)

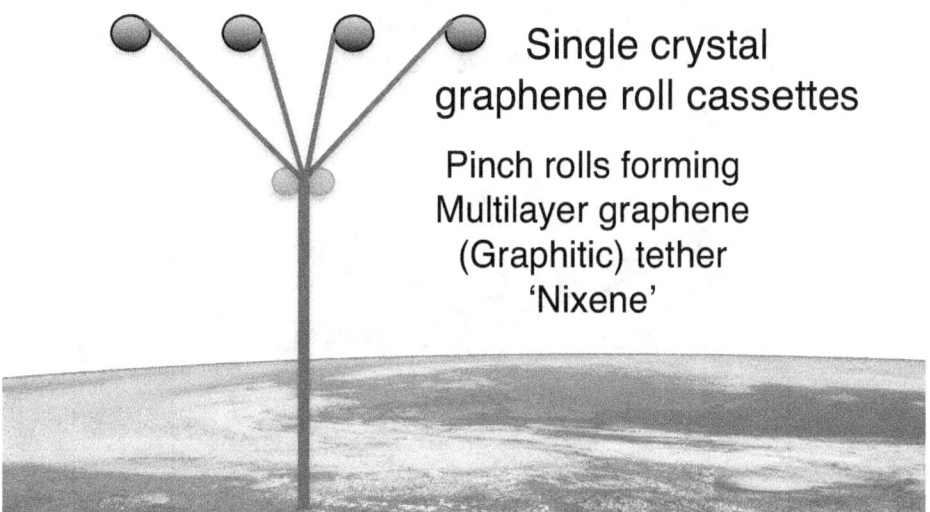

Single crystal
graphene roll cassettes

Pinch rolls forming
Multilayer graphene
(Graphitic) tether
'Nixene'

Figure xx, Combining Tether Layers in orbit
(Slide from Adrian Nixon's presentation webinar June 2019)

Learning to extract best performance out of imperfect materials is a common engineering problem. Rare indeed is the design where all constraints and criteria are fully satisfied by a single solution. Two challenges that separate us from a current tether design become clear. The first, a challenge of assembly—how do we ensure uniform load distribution in our material, so that we can bring the nano-scale properties up to our macro-scale application. The second, a challenge of production—how to scale existing processes up to produce the volume needed. Neither of these challenges requires fundamentally new science or engineering. They require continued application of existing knowledge and skills. Based upon these conclusions, a number of recommendations can be made - the primary one is to encourage and support specific strength material development with the purpose of making them long enough and strong enough for Space Elevator tethers.[58]

C.5   How will the Multi-Stage Space Elevator concept enable early start?

A Multi-stage Space Elevator concept has been developed out of several known technologies with experimentation and operational experiences. The idea behind this creates a platform at an altitude (from 40 to 15,000 km) using a technique developed previously through electromagnetic acceleration of bolts for momentum exchange at the top of the support infrastructure. This building up from the bottom enables a tradeoff with strength of materials for the concept of hanging from GEO altitude. This

[58] Swan, P., David Raitt, John Knapman, Akira Tsuchida, Michael Fitzgerald, Yoji Ishikawa, Road to the Space Elevator Era, Virginia Edition Publishing Company, Science Deck (2019) ISBN-19: 978-0-9913370-3-3

concept has the potential of enabling space elevators earlier than previous plans while relying on known transportation technologies and approaches. As we make progress with the multi-stage space elevator and with materials, we look to a convergence so that the upward reach of the vertical structure meets the downward reach of the tether. (Figure 1)

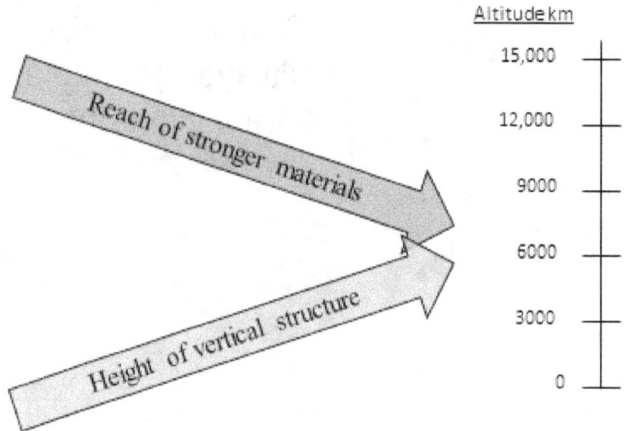

Figure 1 Technology convergence

A good way of looking at progress in development of strong, long, light materials is to see how far they can reach down towards Earth from the geosynchronous altitude (GEO) of 35,850 km. If we had access to the very strong material envisaged by Edwards and others, we would not need the complexity of multiple stages with dynamic support. However, the present work shows that a stable multi-stage structure is possible and could potentially be engineered to use small low-cost components. ISEC Study Report, Multi-Stage .... (2018) has looked at this possibility and there is current research and engineering testing being conducted. One objective in the development of this technology is to implement space elevators earlier than possible with expected strength of materials for the tethers.

C.6   Do we have a "gold standard" for simulations?

Executive Summary[59]
As with all large, modern engineering projects, detailed computer simulations of the space elevator will be essential during its design, construction and operational phases. Within the context of these phases, this study enumerated 14 use cases which

---

[59] Wright, D, S. Avery, J. Knapman, M. Lades, P. Roubekas, P. Swan, "Design Considerations for a Software Space Elevator Simulator," ISEC Position Paper 2017-1.

the simulation software must address, ranging from 3D dynamics and electrodynamics calculations of space elevator motion, to the effects of payload capture and release at various points along the tether, to the effects of friction arising from the interaction of the space elevator climber with the tether. Proceeding from these use cases, requirements were imposed on the software design and an outline for its development was sketched. A central part of the design is a general math and physics platform which can perform the many calculations required. The study team reviewed seven such platforms and chose Mathematica as the one most likely to meet the needs of the simulation. To maintain an open-source option, SageMath was chosen as an alternative math/physics platform. Applications specific to the space elevator simulation will be built on top of these platforms. The simulation software must be developed using modern, best programming practices, and employing Model-View-Controller (MVC) design so that all but a few of the many details of particular space elevator applications are hidden from the user. The simulation must also be modular and flexible enough to evolve with the changing needs of its users. Finally, the software must be made available to a variety of users through various distributed computing technologies such as the cloud. Security issues must be addressed throughout the design and implementation of the software and maintenance will require periodic upgrades and regular testing.

Based on these findings, the study team made 11 recommendations concerning the space elevator simulator. The major ones are:
. A software space elevator simulator should be developed.
. It should be based on Mathematica and SageMath.
. It should be professionally developed and maintained.
. Its development should be funded by a crowd-funding campaign.

C.7   Assured Space Debris Survivability?

Inside this domain of orbital debris, The Space Elevator is a Catalyst for Change! Space debris is expected to be a part of space operations for an extended period in this century. The real mitigation approach is the establishment of policy and actions that will prevent, and extensively reduce, creation of debris in the first place. The Space Elevator must become a catalyst to instigate more aggressive and active removal and mitigation of the space debris challenge. This position paper is International Space Elevator Consortium's (ISEC) attempt to document an approach to mitigate the Space Elevator mission impact and identify safety issues when space debris threatens the Space Elevator's tether. We strongly suggest that other organizations and activities confront this issue also. ISEC believes that debris mitigation concepts will be built, operating, and thriving well before the Space Elevator Transportation System reaches operational status. To that end, this paper serves as the initial characterization of how the Elevator can allocate the needed performance to debris mitigation system then available. That system would serve as Sentry; capturing, destroying, and / or removing the debris threat.

The topic of discussion within ISEC at the present time on the space debris problem is how to work with others and develop programmatic and engineering solutions. Currently, the ISEC leadership is working within the following topics[60]:

- Debris alert ➔ warning needs
- Debris sizing ➔ as a threat variant
- Space Elevator Tether Movement➔ passive defense
- The Sentry System ➔ an architecture adjunct
- System Recovery ➔ post debris-event actions
- Improving the Baseline ➔ configurations to enhance mitigation

The entire matter will remain active in the space elevator systems risk management program. Some activities will be essential for smooth operations.

- Direct, real time coordination will be established with the military Combined Space Operations Control Center [CSpOC]
- There will be a "Debris Assessment Chair" at the Space Elevator Operations Center with unrestrained involvement in daily operations planning and execution.
- As might be necessary, improved Debris Mitigation will be a mandatory inclusion in all system calls for improvement --- (See Architecture Note #29 "Call for Improvement" Policy.)

This position paper will address these topics inside a summary of the problem and a discussion of a global approach towards the future for mitigation of the space debris challenge.

Space Debris is a Manageable Challenge for Space Elevators

The International Space Elevator Consortium's (ISEC) position has been well documented and discussed. The space elevator discussions about space debris were initiated in the 2010 ISEC Study Report, "Space Elevator Survivability, Space Debris Mitigation," after a full year of analyses by space debris and space systems experts. Since then, there have been events that have increased the growth of space debris. This 2020 report has taken a look at the situation and extrapolated across the arena to arrive at some preliminary results. The numbers were calculated for the present (2019 tracked debris data), compared to the past (2010 data), and the future (2030 estimates with projections of new constellations of satellites). The approach, as discussed in the 2010 space debris report, is one where the volume of the space around the Earth is shown to have a density of debris related to altitude zones. That report breaks out the zones, analyzes the information and comes up with conclusions.

---

[60] Fitzgerald, M., "Space Elevator Architecture's Architecture Note #25, Debris Mitigation Roles, ISEC note, www.isec.org, March 2019.

The collision probability analyses are linear with respect to numbers of debris within the volume occupied by 100,000 km of one-meter wide tether. The real concern focuses on high debris density regions with identified zones between 200 and 2000 km altitude. The report takes the density numbers, extrapolates the probabilities of collisions and comes up with conclusions. The Executive Summary of 2010 Report stated: "To assess the risk to a space elevator, we have used methodology from the 2001 International Academy of Astronautics (IAA) Position Paper on Orbital Debris[61]:

The probability (PC) that two items will collide in orbit is a function of the spatial density (SPD) of orbiting objects in a region, the average relative velocity (VR) between the objects in that region, the collision cross section (XC) of the scenario being considered, and the time (T) the object at risk is in the given region."

$$PC = 1 - e^{(-VR \times SPD \times XC \times T)}$$

Using this formula, we calculate the Probability of Collision for Low Earth Orbit (LEO), Medium Earth Orbit (MEO), and Geosynchronous Orbit (GEO). Our focus is on LEO -- as fully two thirds of the threatening objects are in the 200-2000 km (LEO) regime. Our analyses show: The threat from Space Debris can be reduced to manageable levels with relatively modest design and operational "fixes."

With the discussions and calculations across three decades, the conclusion stays the same: for time periods - 2010, 2019 and 2030.

"Space debris mitigation is an engineering problem with definable quantities such as density of debris and lengths/widths of targets. With proper knowledge and good operational procedures, the threat of space debris is not a show-stopper by any means. However, mitigation approaches must be accepted and implemented robustly to ensure that engineering problems do not become a catastrophic failure event."

With the realization that there is much to do in architectural and engineering approaches to space debris mitigation, the following concepts have been assessed as first approximations:

1. Architectural and Engineering Design Inputs. (Multi-leg design, Designing the tether itself to survive the small debris hits, Include a repair tether climber that mends small holes or rips in the tether, and Support operational approaches shown below.)

---

[61] 2001 Position Paper On Orbital Debris, International Academy of Astronautics, 24.11.2000. download for free from www.isec.org

2. Operational Approaches. (Passive Approaches for Debris Mitigation: multi-leg design, varying tether design by altitude, and multiple parallel tethers for greater carrying capacity, and Active Approach: tether movement upon demand, on-orbit Sentry Satellite System, and approach for recovery from tether sever.)

3. Collaboration with Others. (Establish co-operations with Space Traffic Management organizations, Coordinate with owners of space assets (derelicts) at GEO, Coordinate with organizations who will remove space debris, and Establish operational procedures to receive timely warnings and then respond to them)

4. Timely Debris Alert & Warning. (ISEC foresees a close and interactive communication with the military Combined Space Operations Control Center; known familiarly as CSpOC. CSpOC is responsible for tracking the thousands of debris pieces and providing the orbital parameters of those pieces to operating space users. In addition, commercial capabilities have emerged which offer forming and formatting that information; operationally satisfying their commercial customers. Projecting future collisions is an important portion of the tasking for CSpOC, enabling timely warning of the predicted conjunctions to be sent to the space elevator operations center. This timely warning should enable actions to move portions of the tether to avoid those predicted conjunctions. ISEC expects that the space elevator system operator will be able to depend on a warning forecast at 72 hours, of a convergence / close approach to a Space Elevator tether location. The Space Elevator team expects the CSpOC team will hold a position as the Debris Mitigation chair in the Space Elevator Operations Center.)

The Space Debris threat can be reduced to a manageable challenge!!

Appendix D -   Studies ISEC & IAA

The International Space Elevator Consortium conducts year long studies on important topics related to the future development. There are usually 4 to 10 professionals conducting research and submitting chapters for each study. Each of these studies is presented at a International Space Elevator Conference inside a workshop looking for greater diversity of ideas. The resulting study reports reflect their work.

Year	Title
in work	Beneficial Environmental Impacts of the Space Elevator
in work	Galactic Harbour Interplanetary Mission Support
2020	Today's Space Elevator Assured Survivability Approach Space Debris
2019	Today's Space Elevator, Status as of Fall 2019
2018	Design Considerations for a Multi-Stage Space Elevator
2017	Design Considerations for a Software Space Elevator Simulator
2016	Design Considerations for Space Elevator Apex Anchor and GEO Node
2015	Design Considerations for Space Elevator Earth Port
2014	Space Elevator Architectures and Roadmaps
2013	Design Considerations for Space Elevator Tether Climbers
2012	Space Elevator Concept of Operations
2010	Space Elevator Survivability, Space Debris Mitigation

Completed studies can be downloaded as pdf's from www.isec.org and they can be purchased (less than $10 mostly) from www.lulu.com.

Process for ISEC topic development and study process and two recent examples of the ISEC year long process:

- ISEC BOard picks critical topic to be assessed (Aug time period)
- Statement of problem is released with call for contributors (Aug-Nov)
- Team formed and study approach approved (Dec)
- Initial Chapters are submitted as draft (Mar-June)
- Workshop at ISEC Conference or webinar to brainstorm after draft is presented(Aug)
- Final Draft goes to red team
- ISEC year-long study produces final report for pdf download at website and hard copy purchase at www.lulu.com (or amazon.com)

Today's Space Elevator

In the last year, the International Space Elevator

International Space Elevator Consortium   ISEC Position Paper # 2019-1

Today's Space Elevator

Space Elevator Matures into the Galactic Harbour

A Primer for Progress in Space Elevator Development

Peter Swan, Ph.D.
Michael Fitzgerald

Consortium assessed that the basic technological needs for the space elevator can be met with current capabilities: and, each segment of the Space Elevator Transportation System is ready for testing leading to engineering validation. Because of the availability of a new material as a potential Space Elevator tether, the community strongly believes that a Space Elevator will be initiated in the near term. Included are a lexicon of space elevator terms, over 750 references in the bibliography, short descriptions of eight ISEC year-long studies and two IAA 4-year studies on space elevators, as well as a summary of over 20 Architectural Notes covering the development of space elevator technologies. This one document can bring the reader up to speed of the whole space elevator community across policy, technologies, developmental phases, management, and testing progress. Download it now from the www.isec.org - or buy it on www.lulu.com.

Today's Space Elevator Assured Survivability Approach for Space Debris

The International Space Elevator Consortium's (ISEC) position has been well documented and discussed. The space elevator activities about space debris were initiated in the 2010 ISEC Study Report, "Space Elevator Survivability, Space Debris Mitigation" after a full year of analyses by space debris and space systems experts. Since then, there have been events that have increased the growth of space debris. This 2020 report has taken a look at the situation and extrapolated across the arena to arrive at some preliminary results. The numbers were calculated for the present (2019 tracked debris data), compared to the past (2010 data), and the future (2030 estimates with projections of new satellite constellations). The approach, as discussed in the 2010 space debris report, is one where the volume of space around the Earth is shown to have a density of debris related to altitude zones. With discussions and calculations across three decades, the conclusion stays the same: for time periods - 2010, 2019 and 2030. "Space debris mitigation is an engineering problem with definable quantities such as density of debris and lengths/widths of targets. With proper knowledge and good operational procedures, ... space debris is not a show-stopper by any means. However, mitigation approaches must be accepted and implemented robustly."

Appendix E -   Additional Studies

The international Academy of Astronautics has completed two studies on the space elevator topic, each with over 40 global space and space elevator experts. Their reports are published after the 3 to 4 year process.

Year	Title
2019	The Road to the Space Elevator Era
2014	Space Elevators: An Assessment of the Technological Feasibility and the Way Forward
IAA	International Academy of Astronautics - sponsor of study www.iaaweb.org - Virginia Edition Publishing Company, Heinlein Prize Trust https://www.heinleinbooks.com/book-store
2013	The Obayashi Corporation conducted a major study on space elevator design with published results. Dr. Ishikawa,

The Obayashi Corporation has conducted a study of "how to build a space elevator." Their results are quite comprehensive and stimulate considerations for the future. Their inputs have been contributed towards both the IAA studies and the ISEC activities.

"Space Elevators: An Assessment of the Technological Feasibility and the Way Forward." 2015

What are the questions for this study report? This report addresses the simple and complex issues that have been identified through the development of space elevator concepts over the last decade. The report begins with a summary of those ideas in Edwards' and Westling's book "The Space Elevator" (2003). Out of these beginnings has risen a worldwide cadre focused upon their areas of expertise as applied to space elevator development and operational infrastructure. The report answers some basic questions about the feasibility of a space elevator infrastructure. A preview of the main questions and answers shows the depth and breadth of this Cosmic Study.

What is a space elevator? A space elevator is a system for lifting payloads, and eventually people, from the Earth's surface into space. The one under consideration in this report consists of a tether 100,000km long balanced about a node in geosynchronous orbit (GEO) and reaching down to an anchor point on Earth. Electrically powered spacecraft, called tether climbers, travel up or down the tether at far lower costs [currently projected at $500/kg] than using rockets. In addition, the service the space elevator provides is a cargo capacity/throughput of two orders of magnitude larger than present rockets, with tremendously kinder environmental effects, and a miniscule potential for future space debris. Tether climbers can continue to the apex anchor – the point at 100,000km altitude – where their speed is sufficient for direct interplanetary travel.

Why a space elevator? The value and benefit of developing a space elevator infrastructure is even greater than earlier estimates, as it will change our approach to operations in space. Low cost, safe, reliable and flexible delivery of payloads to Geosynchronous Earth Orbit (GEO) and beyond could create an "off-planet" environment filled with opportunities ranging from commercial space systems to exploration of the solar system. Daily initiation of 20 metric ton climbers, safe delivery to GEO and beyond, and a projected price of $500 per kg, will open up the solar system and lead to many new commercial ventures. In addition, the radical change from chemical rockets and the low risk approach of climbing vertically at reasonable speeds will greatly reduce two major hazards that are dominant today: 1) the environmentally friendly, electrically driven, motors will have almost no hazardous material polluting the atmosphere, and 2) this delivery technique does not create orbital debris, especially in Low Earth Orbit (LEO).

Another major benefit will be in supporting human exploration. The first ten years will enable massive movement of equipment to GEO and beyond. Human exploration can leverage this tremendous capability by assembling large spacecraft at GEO with massive fuel loads delivered at $500/kg. After ten years of operations, humans should also be riding to GEO.

The benefits for humanity on Earth can be phenomenal. The ability to inexpensively deliver large quantities to orbit will enable capabilities stimulating an Earth renaissance. The facility to provide power to any location on the surface [space solar power satellites] will enable development across the world. Several examples are that Africa could skip the last century of wires while the outback of countries like India or China would not have to burn coal and the Amazon region could retain more of its rain forests. In addition, the increase in communications and Earth resource satellites will remake the emergency warning systems of the world. Some intractable problems on the Earth's surface would also have solutions, such as the safe and secure delivery – and thus disposal - of nuclear waste to solar orbit.

Can it be done? The authors recognize that the whole project, especially the projected price per kilo, is dependent upon a strong, lightweight material that will enable the space elevator tether. The principal issue is material produceability at the strength, length and perfection needed to enable a 100,000km long tether. Almost all other issues surrounding each of the major segments have either been resolved in space before or are close to being space ready today. Only the tether material is at a high technological risk at this time. Chapter 3 goes into projections of material growth and increase in capabilities showing their potential with a good prospect of suitable material becoming available by the 2020s.

How would all the elements fit together to create a system of systems? Each of the early chapters addresses one of the major elements of space elevator infrastructure. As the study progresses, the reader moves from tether material to individual segments to systems level analyses. This sequence illustrates the parts of a space elevator infrastructure and then shows the operational view as it all fits together. In addition, in the market and financial chapters, the development of future space markets are projected with their funding profiles for the next 40 years.

What are the technical feasibilities of major space elevator elements? Each of the individual chapters describes major segments of the space elevator and discusses NASA Technical Readiness Levels and Risk Management trades to ensure the technical feasibilities can be assessed. The space elevator roadmaps show the approach from the current year [2013] to operational time periods. A factor for the future of space elevator infrastructure is the majority of components, subsystems, and segments have been developed before as components of other space systems [except for the tether material]. This leverage of 50 years' experience is invaluable and will enable development of space elevator segments in a timely manner.

The conclusions from this study fall into a few distinct categories:
- **Legal:** The space elevator can be accomplished within today's arena
- **Technology:** Its inherent strengths will improve the environment and reduce space debris in LEO and beyond. It can be accomplished with today's projection of where materials science and solar array efficiencies are headed. The critical capability improvement is in the space elevator tether materials, currently projected to achieve the necessary strength to weight ratio in the next 20 years. The space elevator will open up human spaceflight and decrease space

debris and environmental impacts.

- **Business:** This mega-project will be successful for investors with a positive return on investment within 10 years after erection is complete.
- **Cultural:** This project will drive a renaissance on the surface of the Earth with its solutions to key problems, stimulation of travel throughout the solar system, with inexpensive and routine access to GEO and beyond.

Cosmic Study Result The authors have come to believe that the operation of a space elevator infrastructure will lead to a "game changing" experience in the space world. Each of the authors considers that the space elevator can be developed when the tether material is mature enough. Our final assessments are:

A Space Elevator appears feasible, with the realization that risks must be mitigated through technological progress. and, A Space Elevator infrastructure will be achievable through a major global enterprise.

Road to
Space Elevator Era

Editors:
Peter A. Swan
David I. Raitt
John M. Knapman
Akira Tsuchida
Michael Fitzgerald

International Academy of Astronautics

"Road to Space Elevator Era." 2019
This study report summarizes the assessment of the space elevator as of the summer of 2018. The encouraging aspect is that the space elevator community has been reinvigorated and is pulling together experiments and test programs to push the technology along the path to readiness. Several of these break through are the ones we were searching for after completion of the first IAA study. We see the way forward! The global needs for a space elevator are remarkable. When the price to geosynchronous orbit is lowered to one hundredth of the price of launching by rocket, the whole situation changes as to access to orbit. However, the real strengths are not only price but massive movement and other characteristics such as routine, daily, safe and no shake-rattle-roll of launch. The environmentally friendly lifts will be an important aspect of implementing space elevators vs. rockets in the long run. One key to recognize is that we move from individual events to continuous operations of an infrastructure with the space elevator. We would move to a system with the costs representing recurring expenses, not replacement costs. The concept is to move to a "bridge to space," not a system of individual rocket launches.   The question on the table is "are we actually on the road to a space elevator?" The study

answers that question in a positive manner. Yes, we are on the road to the space elevator era!

This study report assesses the global needs for a space elevator and then lays out functional requirements that can lead to technological needs and identification of processes for development. The development of the needs and requirements lead to the chapters assessing the critical technologies and then recommends the risk reduction approaches for segment and system level verification. In addition, Chapter Six defines validation testing for the customer, as a milestone towards total project funding.

As the goals of the study were approached systematically during the study, the results were presented in the form of conclusions and recommendations. When one looks at all the various technologies and where they are in the technology readiness level (TRL) evaluations common to NASA projects, the team has the following conclusions:

- The Earth Port is buildable with today's available technologies and engineering expertise.
- The Headquarters and Operations Centers are buildable today.
- The tether climber is so similar to a normal satellite design of today that there is no real technological or engineering challenge; except the interface with the tether material. As there is a lack of information of the chosen material for the tether, some engineering must be resolved at a later time.
- The GEO Node and GEO Region technologies are understandable and not an issue during development.
- The Apex Anchor will be a challenge as its role is key to the building of the space elevator, but not an engineering and technological issue.
- However, the tether material is the pacing item for the development of the space elevator. Currently, there are three viable materials that could grow into the needed strong-enough and long-enough material for a space elevator: carbon nanotubes, boron nitride nanotubes, and continuous growth graphene. The community waits for those materials to mature to the level that can be implemented into a space elevator tether 100,000km long and strong enough to support its own weight plus multiple tether climbers against the pull of gravity. At 100 million metres long, a space elevator tether is a major engineering challenge. Recent investigations explored the possibility for making single crystal graphene by a continuous process using liquid metal. Making this a viable practical manufacturing process will be a significant effort over a period of years and probably many millions of dollars. However, such a process would create graphene products for many multi-billion dollar markets on the way to making the tether material. For this reason there is a credible return on investment case for manufacturing the material in practise. This means it is highly possible that continuous single crystal graphene will be manufactured in the coming years and this material should be considered in

any forward thinking about a space elevator tether.

- Learning to extract best performance reel-out of imperfect materials is a common engineering problem; rare indeed is the design where all constraints and criteria are fully satisfied by a single solution. Two challenges that separate us from a tether become clear. The first, a challenge of assembly—how do we ensure uniform load distribution in our material, so that we can bring the nanoscale properties up to our macroscale application. The second, a challenge of production—how to scale existing processes up to produce the volume needed. Neither of these challenges requires fundamentally new science or engineering. They require continued application of existing knowledge and skills.

Based on these conclusions, a number of recommendations can be made - the primary one of which is to encourage and support specific strength material development with the purpose of making them long enough and strong enough for space elevator tethers.   The essence of the whole study report is that a broad group of space professionals gathered together and assessed the status of the space elevator development. Each of them contributed their expertise and then came to similar conclusions about the space elevator progress. It should be noted that, even though many of the references have several authors, for convenience in the text only the first named author is given.

Space Elevators seem feasible [reinforces IAA's 2013 study conclusion],
and
Space Elevators are now ready to go into segment level testing
[reinforces IAA's 2019 study conclusions].

Appendix F -   Description of ISEC

The International Space Elevator Consortium (ISEC) is composed of individuals and organizations from around the world who share a vision of humanity in space. www.isec.org has much more on the topic.

OUR VISION:

- A world with inexpensive, safe, routine, environmentally friendly, and efficient access to space for the benefit of all mankind.

OUR MISSION:

- The International Space Elevator Consortium (ISEC) promotes the development, construction and operation of a Space Elevator (SE) Infrastructure as a revolutionary and efficient way to space for all humanity. ISEC is made up of individuals and organizations from all around the world who share a vision of mankind in space.

What We Do:

- Provide technical leadership promoting development, construction, and operation of space elevator infrastructures.

- Become the "go to" organization for all things space elevator.

- Energize and stimulate the public and the space community to support a space elevator for low-cost access to space.

- Stimulate STEM* activities while supporting educational gatherings, meetings, workshops, classes, and other similar events to carry out this mission.

A BRIEF HISTORY

The idea for an organization like ISEC had been discussed for years; but, it wasn't until the Space Elevator Conference in Redmond, Washington, in July of 2008, that things became serious. Interest and enthusiasm for the SE had reached an all-time peak; and, with Space Elevator conferences upcoming in both Europe and Japan, it felt that this was the time to formalize an international organization. An initial set of Directors and officers were elected and they immediately began the difficult task of unifying the disparate efforts of Space Elevator supporters world-wide. ISEC's first Strategic Plan was adopted in January of 2010 and it is now the driving force behind ISEC's efforts. This Strategic Plan calls for adopting a yearly theme (for 2010 it was "Space Elevator Survivability – Space Debris Mitigation") which focuses ISEC activities. Because of the common goals and hopes for the future of mankind, off-planet, ISEC became an

Affiliate organization with the National Space Society, in August of 2013. (See History page on the website)

OUR APPROACH

ISEC's activities are pushing the concept of space elevators forward. These cross all the disciplines and encourage people from around the world to participate. The following activities are being accomplished in parallel;

- Yearly Conference – International space elevator conferences were initiated by Dr. Edwards in the Seattle area in 2002. Follow-on conferences were in Santa Fe [2003], Washington DC [2004], Albuquerque [2005/6 –smaller sessions], and Seattle [2008 to the present]. At each of these conferences, there were many discussions across the whole arena of space elevators with remarkable concepts and presentations. The current conferences [beginning in 2012] have been sponsored by Microsoft, Seattle Museum of Flight, Space Elevator Blog, Leeward Space Foundation, and ISEC. The last several conferences were co-sponsored by and held in the Seattle Museum of Flight.

- Year Long Studies – ISEC sponsors research into a focused topic each year to ensure progress in a discipline within the space elevator project. The first was in 2010 and evaluated the threat of space debris. The second study, and resulting report focused on the Concept of Operations of space elevators. The products from these studies are reports that are published to show progress in the development of space elevators. [see free pdf's under "Studies"]

- International Cooperation – The ISEC supports many activities around the globe to ensure that space elevators keep progressing towards a developmental program. International activities include coordinating with the two other major societies focusing on space elevators; Japanese Space Elevator Association & EuroSpaceward. In addition, ISEC supports symposia and presentations at the International Academy of Astronautics and the International Astronautical Federation Congress each year.

- Competitions – ISEC has a history of actively supporting competitions that push technologies in the area of space elevators. The initial activities were centered on NASA's Centennial Challenges called "Elevator: 2010." Inside this were two specific challenges: Tether Challenge and Beam Power Challenge. The highlight was when Laser Motive won $900,000, in 2009, as they reached one kilometer in altitude racing other teams up a tether suspended from a helicopter. There were also multiple competitions where different strengths of materials were tested going for a NASA prize – with no winners. In addition, ISEC supports educational efforts of various organizations, such as the LEGO space elevator climb competition, at our Seattle conference. Competitions have also been conducted in both Japan and Europe.

- Publications – ISEC publishes a monthly e-Newsletter and yearly study reports, to help spread information about Space Elevators.

- Reference Material – See Library, including a reference database of Space Elevator related papers and publications.

- Outreach – People need to be made aware of the idea of a Space Elevator. Our Outreach activity is responsible for providing the blueprint to reach societal, governmental, educational, and media institutions and exposes them to the need for a Space Elevator. ISEC members are readily available to speak at conferences and other public events in support of the Space Elevator. In addition to our monthly e-Newsletter, we are also on Facebook, Linked In, and Twitter.

- Legal – The Space Elevator is going to break new legal ground. International cooperation must be sought. Insurability will be a requirement. Legal activities encompass the legal foundation of the Space Elevator – international, maritime, air, and space law. Also, there will be interest within IP, liability, and commerce law. Starting work on the legal foundation well in advance will result in a more rational product – legal opinions that pre-date a Space Elevator.

Appendix G - Lexicon of Terms

After many meetings and discussions with players from around the world, a layout of a space elevator is shown below with a set of terms and their explanation following:

Apex Anchor Node & Region	LEO Gate	Earth Port and Region
Mars Gate	Lunar Gravity Center	- Earth Terminus
Moon Gate	Mars Gravity Center	- Floating Operations Platform
GEO Node and Region	Tether Climbers	Headquarters and Primary Operations Center

The following pages show the basics of the space elevator with definition of terms and figures that help define our concepts of space elevators. The following sections are expanded upon: SE Lexicon [with figure], Terminology Table, ISEC's Galactic Harbour Strategic Approach [with figure], SE Regions, SE Developmental Sequences, and our destinations.

Figure: Space Elevator System

Table of Suggested Terminology

Terminology	*Explanation*
Access City	Earth Port Access City will be the principle location where the majority of supplies/payloads depart from in route to the Earth Port. It should be the location for the HQ/POC and within 2,500 kilometers of Earth Port Region.
Apex Anchor	A complex of activity is located at the end of the Space Elevator providing counterweight stability for the space elevator as a large end mass. Attached at the end of the tether will be a complex of Apex Anchor elements such as; reel-in/reel-out capability, thrusters to maintain stability, command and control elements, etc.. Release from AA enables "free flights to Mars and beyond" rapidly and daily. [Note: nothing stays at that altitude unless attached to a tether]
Apex Anchor Region	The region around the Apex Anchor is defined by the amount of motion expected at the full extension of the tether. The region is the volume swept out by the end of the tether during normal operations. When two or more space elevators are operating together, the region spreads to the volume between.
Boron-Nitride Nanotube (BNNT)	High Tensile Strength material under development
Boron Nitride "white graphene" Hexagonal	Hexagonal Boron Nitride is similar to graphene hexagonal layers of alternating atoms of boron and nitrogen and nearly as strong as graphene.
Capability On Ramps leading to FOC	Time after IOC when new businesses / capabilities are added to system [7th sequence step]
Carbon Nanotube (CNT)	High Tensile Strength material under development
Climbers [Tether Climbers]	Vehicle able to climb or lower itself on the tether
Deployment	Releasing the tether from the GEO construction up and or down during the initial phase of construction
Earth Anchor (Tether Terminus)	Earth Terminus for space elevator
Earth Port	A complex located at the Earth terminus of the tether to support its functions. These mission elements are spread out within the Earth Port Region. When there are two or more termini of tethers, the Earth Port reaches across the region and is considered one Earth Port.
Earth Port Region	The volumetric region around each Earth Port to include a space elevator column for each tether and the space between multiple tethers when they operate together. The Earth Port Region will include the vertical volume through the

atmosphere up to where the space elevator tether climbers start operations in the vacuum and down to the ocean floor.

Floating Operations Platform	The Op's Center for the activities at the Earth Port or Earth Terminus
Full (Final) Operational Capability (FOC)	Design for full capability of the space elevator [8th sequence step]
Galactic Harbour	A new vision of Galactic Harbour architectures has developed. The Galactic Harbour is the combination of the Space Elevator Transportation System and the Space Elevator Enterprise System. The Galactic Harbour will be the volume encompassing the Earth Port while stretching up in a cylindrical shape to include two Space Elevator tethers outwards beyond the Apex Anchor.
GEO Node	The complex of Space Elevator activities positioned in the Space Elevator GEO Region of the Geosynchronous belt [36,000 kms altitude]; directly above the Earth Port. There will be several sub nodes; one for each tether, one for a central main operating platform, one for each "parking lot", and others.
GEO Region	Encompasses all volume swept out by the tether around the Geosynchronous altitude, as well as the orbits of the various support and service spacecraft "assigned" to the GEO Region. When two or more space elevators are operating together, the region includes each and the volume between elevators.
Headquarters and Primary Operations Center [HQ/POC]	Location for the Operations and Business Centers – probably other than at Earth Port – more likely near Space Elevator Access City
Initial Operational Capability (IOC)	A term to describe the time when the space elevator is prepared to operate for commercial profit – robotically [6th sequence step]
International Academy of Astronautics (IAA)	International Association focusing upon space capabilities with approximately 1,000 elected members.
International Space Elevator Consortium (ISEC)	Association whose vision is: A world with inexpensive, safe, routine, and efficient access to space for the benefit of all mankind.
Japanese Space Elevator Association	JSEA handles all the space elevator activities for universities and STEM activities. Also handles the global aspects of space elevators.
Japanese Space Agency (JAXA)	Japanese government organization responsible for space systems and space operations.
Length Overall	Full length of the space elevator, est. from 96,000 to 100,000

	km
LEO Gate	Elliptical release point for LEO – roughly 24,000 kms altitude
Limited Operational Capability	Early utilization of a "starter" tether in parallel with testing and further development [5th sequence step]
Lunar Gate (Moon Gate)	Release Point towards Moon – roughly 47,000 kms altitude
Lunar Gravity Center	Point on Tether with Lunar gravity similarity – 8,900 kms altitude
Marine Node (Earth Port)	Earth Terminus for space elevator
Mars Gate	Release Point to Mars – roughly 57,000 kms altitude
Mars Gravity Center	Point on Tether with Mars gravity similarity – 3,900 kms altitude
Ocean Going Vehicle (OGV)	Vehicle able to travel over the open ocean
Operational Testing	Key developmental phase when checking out capability [4th sequence step]
Pathfinder	In-orbit testing of space elevator with as many segments represented as possible [1st sequence step]
Primary Operations Center (POC)	Center of all activities for the space elevator. Could be distributed or centralized.
Seed Tether [Ribbon]	The initial tether lowered from GEO altitude which would then be built up to become the space elevator tether [2nd sequence step]
Single Crystal Graphene	Single Crystal Graphene is a flat, continuous sheet of sp2 hybridised carbon. We use the term to refer to this advanced material at scales from cm2 to m2. Should be long enough and strong enough for space elevator tethers.
Single String Testing	Single string tests are tests conducted of a selected set of Space Elevator functions; aligned and operating. In early forms, single string testing could be an end-to-end simulation of a segment. Later, hardware is inserted in the string to add realism. Testing the initial tether after deployment would be a key single string test. [3rd sequence step]
Space Elevator Column	The volume swept out during normal operations starting at the Earth Port [a circular area within which it operates] and extending through the GEO Region up to the Apex Region. This column of space will be monitored, restricted, and coordinated with all who wish to transverse the volume.
Tether	100,000 km long woven ribbon of space elevator with sufficient strength to weight ratio to enable an elevator [CNT material probably]
Tether Climbers	Vehicle able to climb or lower itself on the tether, as well as releasing or capturing satellites for transportation or orbital insertion.

Tonne, Metric Tonne   1,000 kg (sometimes incorrectly shown as MT)

ISEC's Galactic Harbour Strategic Approach

One of the principle elements of the International Space Elevator Consortium's (ISEC) action plan towards an operational space elevator is to understand its customer utilization. To fully understand the potential application for commercial ventures on the space elevator, the concept of a Galactic Harbour surfaced. Galactic Harbour represents continuous operations moving customer payloads on multiple space elevators from entry ports to exit ports. These locations would most logically be an Earth Port where customers have their payloads loaded onto space elevators and their release points at multiple altitudes per the desires of the customers. The Galactic Harbour would then be the volume incorporating multiple Earth Ports [on the ocean, with incoming and outgoing ships/helicopters and airplanes] and then stretch up in a cylindrical shape to include tethers and other aspects out to the Apex Anchors.

Galactic Harbour as the Unification of Transportation and Enterprise

Space Elevator Transportation is the "main channel" in the Galactic Harbour	Businesses flourish as a part of the Space Elevator Enterprise System
GEO Node	Business support to Operational Satellites
Earth Port	Power and products delivered to Earth
Apex Region	Interplanetary Efforts within reach
Tether Climbers	Research
Tether System	
HQ/POC	

Our Destinations
- *The Initial Operational Capability (IOC) consists of a system comprised of two space elevators with one Earth Port and two terrestrial terminus, two Apex Anchors each with 100,000 km tethers, multiple tether climbers and a single Headquarters and Primary Operations Center. This system will be capable of moving significant payload tonnage [20 Metric ton - 14 MTs of cargo] to GEO and beyond several times a week from each space elevator.*

- *The Full Operational Capability (FOC) contains two tethers per elevator system (100,000 km strong tether), each with a tether terminus platform inside the Earth Port, GEO Node, Apex Anchor, and with a single Headquarters and*

141

Primary Operations Center. This system will be capable of moving an estimated 100 Metric tons (79 MTs of cargo) to GEO and beyond several times a week (with passengers).

Space Elevator Regions

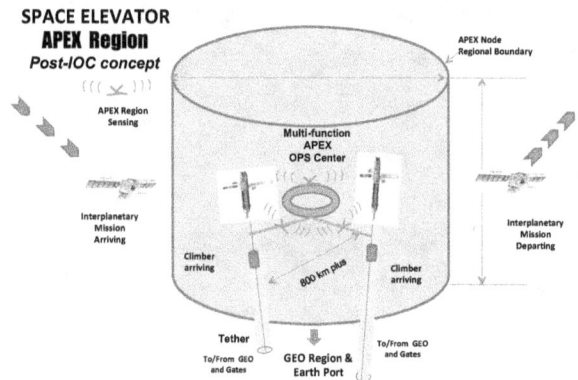

Apex Anchor Region: The region around the Apex Anchor is defined by the amount of motion expected at the full extension of the tether. The region is the volume swept out by the end of the tether during normal operations. When two or more space elevators are operating together, the region spreads to the volume between.

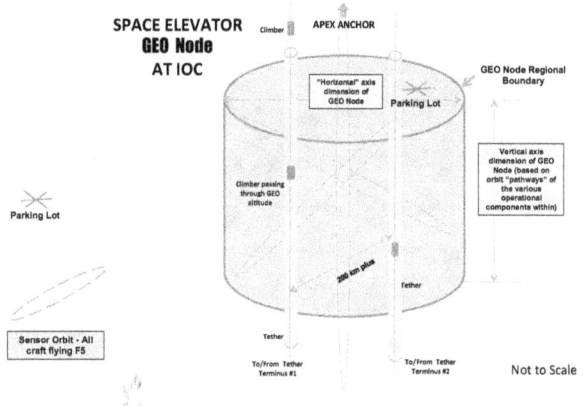

GEO Region: Encompasses all volume swept out by the tether around the Geosynchronous altitude, as well as the orbits of the various support and service spacecraft "assigned" to the GEO Region. When two or more space elevators are operating together, the region includes each and the volume between elevators.

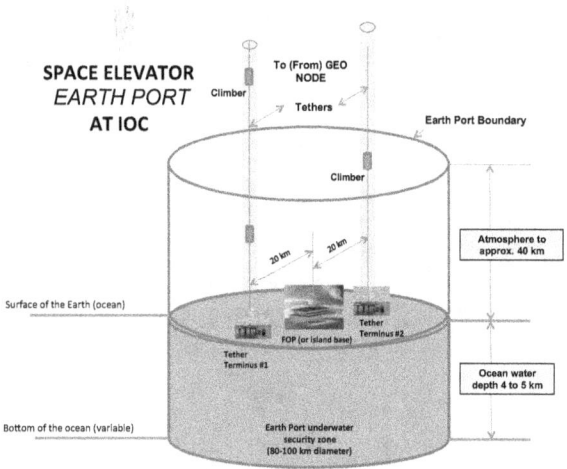

Earth Port Region: The volumetric region around each Earth Port to include a space elevator column for each tether and the space between multiple tethers when they operate together. The Earth Port Region will include the vertical volume through the atmosphere up to where the space elevator tether climbers start operations in the vacuum and down to the ocean floor.

Space Elevator Development Sequence
Setting the stage with a typical life cycle phase schedule for developing space systems. Space Elevators are still, very much, in the Concept Development phase.

Typical Project Life Cycle Phases

Project Life Cycle Phases	Pre Phase A: Concept Study	Phase A: Concept & Technology Development	Phase B: Preliminary Design and Technology Completion	Phase C: Final Design & Fabrication	Phase D: System Assembly, Integration & Test, Launch	Phase E: Operations & Sustainment	Phase F: Closeout
Reviews -Mission		MCR MDR					
Reviews -System		SRR SDR	PDR	CDR	ORR FRR		

Formulation Phase (More Academic level efforts are required)

Implementation Phase (Space Agency, Private sector, Industries, etc.)

Space Elevator Development

Space Elevator On-orbit Assembly, Checkout, and Operations

We are still here.

<Notes>
MCR: Mission Concept Review, MDR: Mission Definition Review, SRR: System Requirements Review, SDR: System Definition Review, PDR: Preliminary Design Review, CDR: Critical Design Review, ORR: Operational Readiness Review, FRR: Flight Readiness Review
(Ref: NPR7123.1A NASA Systems Engineering Processes and Requirements w/Change 1 (11/04/09))

Special Sequence for Development of Initial Space Elevators

1. Pathfinder
2. Seed Tether,
3. Single String Testing
4. Operational Testing,
5. Limited Operational Capability (LOC),
6. Initial Operational Capability (IOC),
7. Capability On Ramps leading to FOC
8. Full Operational Capability (FOC)

Note: it is critical to initiate second tether for backup as soon as practical – in this sequence, Deployment 2 should be right After Single String Testing, with IOC only when two space elevators are up.

Appendix H - Space Policy Directive - 1: An Expansion

Bethesda, MD[62] - In December 2017, The President gave NASA a new mission directive: work with international and commercial partners to refocus exploration efforts on the moon, with an eye to eventually going on to Mars and even beyond. As stated in Space Policy Directive-1 (SPD -1), "The NASA Administrator shall, 'Lead an innovative and sustainable program of exploration with commercial and international partners to enable human expansion across the solar system and to bring back to Earth new knowledge and opportunities. Beginning with missions beyond low-Earth orbit, the United States will lead the return of humans to the Moon for long-term exploration and utilization, followed by human missions to Mars and other destinations.' "

Simply put, NASA's future space exploration campaign includes efforts focused on three core domains: low Earth orbit; lunar orbit/surface; Mars and other deep space objectives. The four primary strategic goals are:
- Transition U.S. human spaceflight in low-Earth orbit to commercial operations, which support NASA and the needs of an emerging private sector market
- Extend long-duration U.S. human spaceflight operations to lunar orbit
- Enable long-term robotic exploration of the Moon
- Enable human exploration of the Moon as preparation for human missions to Mars and beyond

As a result of this change in direction, two low-Earth orbit objectives have changed:
- End direct government support to ISS by 2025 while stimulating commercial activities that utilize and meet NASA's exploration risk mitigation and science requirements
- Generally increase the breadth and depth of commercial and international LEO missions

Primary lunar orbit/surface objectives are:
- Establish a long-term presence in the vicinity of and on the Moon, including science and human exploration advancement, while enabling national and commercial goals
- Study a gateway that could allow for earlier crew expeditions, more science and technology demonstration capabilities and increased room for astronauts to live and work

Mars and other deep space objectives include:
- Advance U.S. leadership at Mars with a rover as a first step toward a sample-return strategy
- Search for past life and demonstrate oxygen production
- Demonstrate a round-trip robotic mission with the historic first launch off another planet and sample return through a lunar gateway
- Develop standards for human long-duration deep space transportation vehicles

In summary, the President has given NASA a very challenging set of space exploration objectives to pursue in the coming decade. Let's see if technology advances and federal budgets will support this level of ambition

---

[62] "Back to the Moon and On to Mars," Launchspace eNewsletter, July 13, 2020.

www.ingramcontent.com/pod-product-compliance
Lightning Source LLC
Chambersburg PA
CBHW081512220526
45467CB00010B/2890